URANUS, NEPTUNE, PLUTO

A longer view

of their history and their movements in the years ahead

Gods of the sky,
the sea,
the underworld

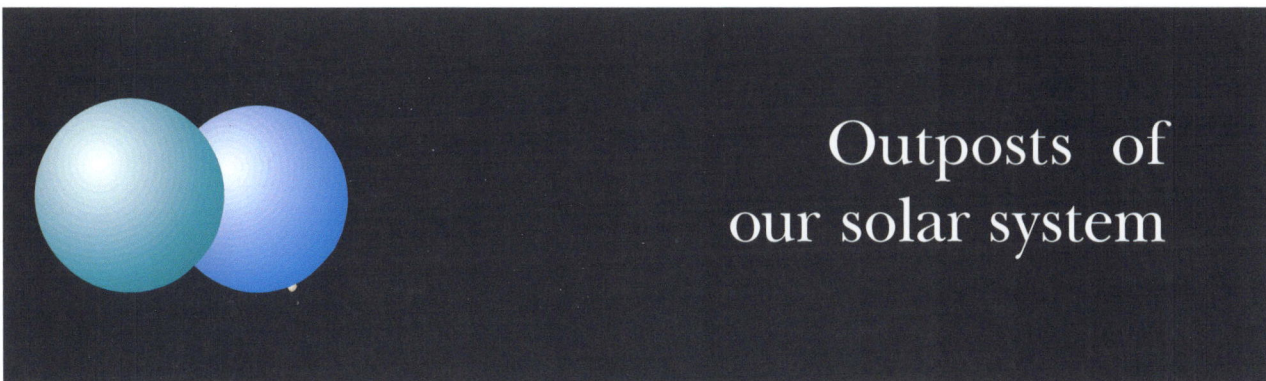

Outposts of
our solar system

Guy Ottewell

ISBN 978-0-934546-75-1

State 5, September 2018

Universal Workshop
www.universalworkshop.com
Greenville, S.C., U.S.A., and Lyme Regis, Dorset, England

Preface

Every time I ground through the process of preparing my annual book, the *Astronomical Calendar*, I wished that instead—

Well, start again. The first issue of that book, for 1974, was a simple product with two pages for each month, put together with drawing tools and typewriter at a time when I knew how to find the stars and planets but knew nothing of mathematics, astrophysics, printing technology, publishing business, or computers. Circumstances caused me to stick to this one among my interests, and the annual books grew large and ever more elaborate, culminating in the last, for 2016.

Each year, or at any rate in the last several years, as I ground through the protracted task, the thought occurred to me that instead of preparing the section on each planet for the year ahead, I would rather be doing it for a stretch of years ahead. Then the chart could show the whole pattern of, say, Jupiter's looping movement through its twelve-year tour of the constellations; and instead of yet again reworking a brief paragraph on why Mercury's appearances are so odd, I would have space for an adequate explanation; and instead of choosing among the auxiliary diagrams I had devised for the movements of Uranus and Neptune, I would have space to include them all.

So this book is the first of what I hope will be a "Longer View" series. The charts trace Uranus, Neptune, and Pluto through at least a dozen years—starting with 2017 because that was the first year for which I didn't print an *Astronomical Calendar*.

Uranus, Neptune, and Pluto are linked in a chain of discovery. Though Pluto is no longer considered the ninth major planet, it is worth including because it is just within the reach of skilled amateur observers.

I am grateful to John Goss, President (2014-2018) of the Astronomical League, for proof-reading this book and saving me from many typos and other minor errors. Any that survive are my own responsibility.

Readers are welcome to send corrections or suggestions (as some have already done) to guy@universalworkshop.com.

4

Contents

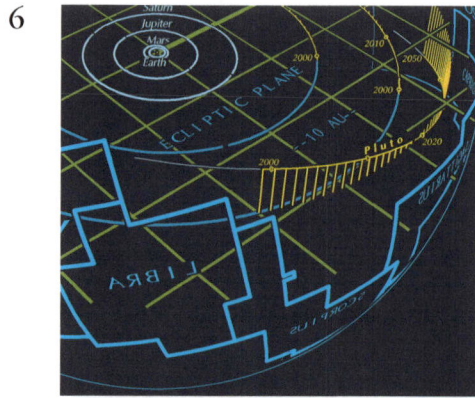

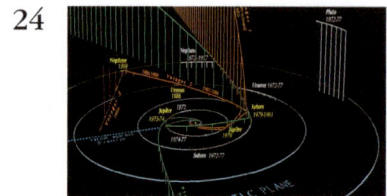

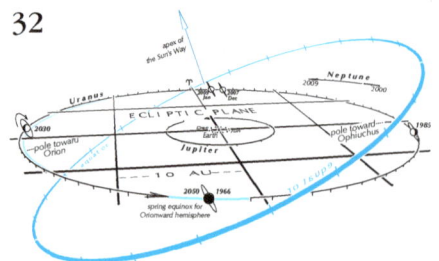

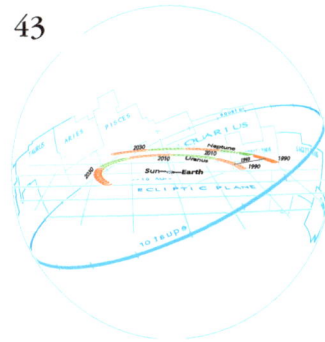

43

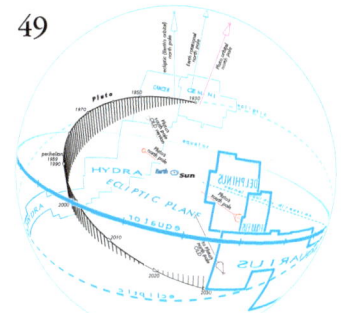

49

54

66

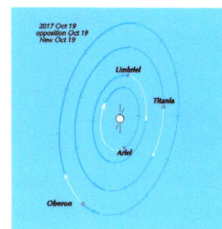

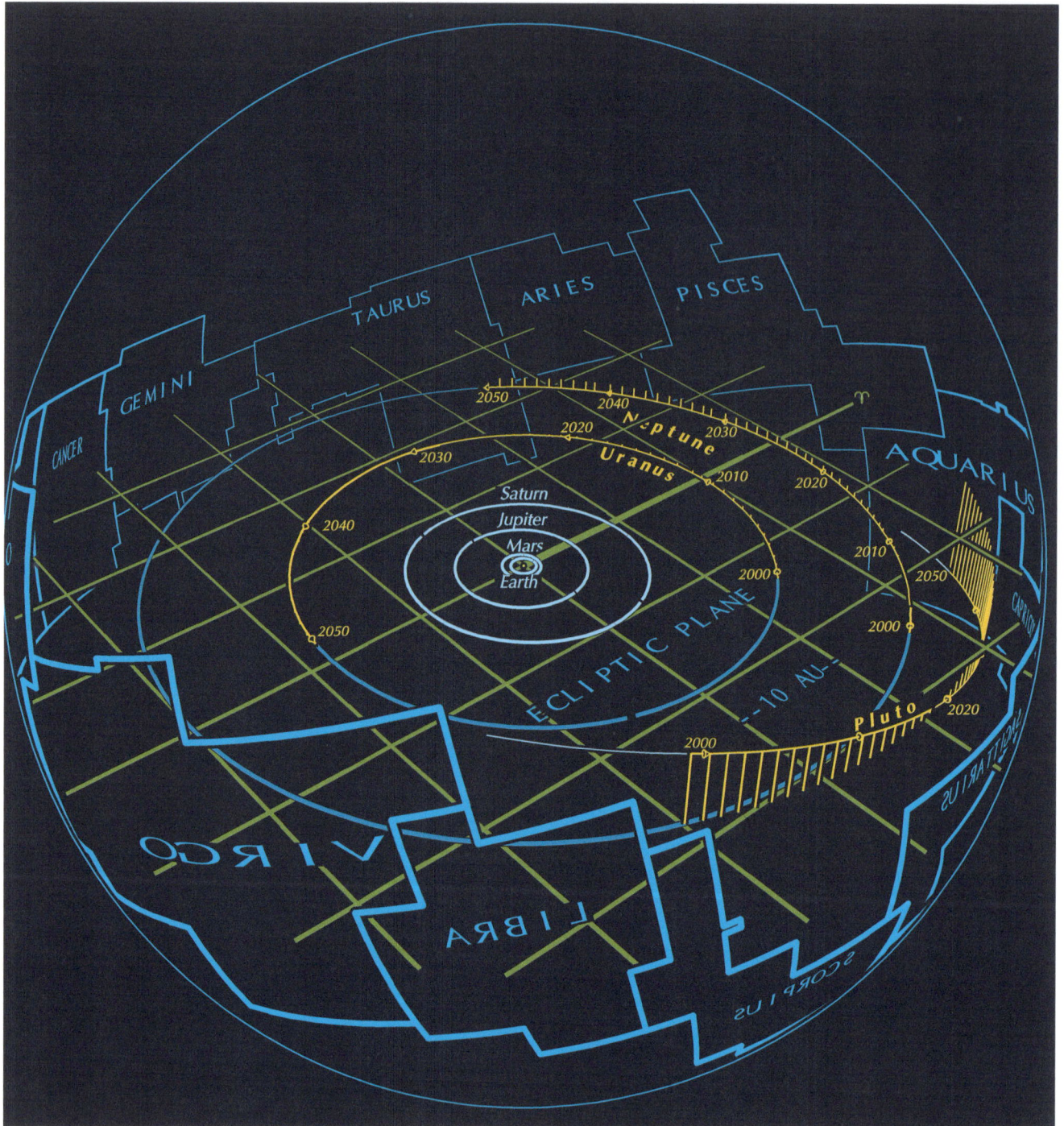

A space view showing the travels of Uranus, Neptune, and Pluto in the first half of the 21st century: 2001 to 2051. Arrowheads mark the ends of round-number years 2010, 2020, etc. At the beginnings of all years are stalks perpendicularly southward or northward to the ecliptic plane.

The Sun is the dot in the middle of the picture. You are looking from a viewpoint 120 astronomical units (Sun-Earth distances) from the Sun, and 35° north of the ecliptic plane. The boundaries of the 12 zodiacal constellations (through which the ecliptic passes) are drawn on an imaginary sphere with radius 40 a.u., centered on the Sun. Grid lines on the ecliptic plane are 10 a.u. apart. The thicker line is the vernal equinox direction, along which Earth sees the Sun at the March equinox; it is the zero point for astronomical measurements, and is marked with the ram's-horns symbol (it was traditionally called the "First Point of Aries," though precession has now moved it into Pisces).

Introduction

Uranus and Neptune were the first "new" planets, added by discovery in the 18th and 19th centuries to the six that had long been known. Their discovery led in the 20th century to that of Pluto, which was at first regarded as the ninth planet but proved to be the first of a swarm of small and even more remote bodies.

Thus the family of major planets came to consist of two divisions: the four smaller inner ones (Mercury, Venus, Earth, Mars) and the four outer giants (Jupiter and Saturn and the somewhat less gigantic Uranus and Neptune). The first four are sometimes called "terrestrial" (Earth-like) and the four outer "Jovian" (Jupiter-like). Another way of classifying the planets other than Earth is that Mercury and Venus are the inferior planets (the word means "lower" in Latin) because they are down nearer to the Sun than we are; Mars and outward are the superior planets.

Uranus, Neptune, and Pluto are enchained by the human story of their discovery. Though below, or in Uranus's case questionably at, the naked-eye limit, they are not too dim to be potentially observable by skilled amateurs with good telescopes.

The story

The ancient planets

Among the myriad lights in the sky, people from early times noticed seven that move, as if alive: the Sun, Moon, Mercury, Venus, Mars, Jupiter, and Saturn. These were called in Greek the *planêtai*, "wanderers," as opposed to the "fixed stars." They were associated with, or thought to be, gods, or at least to have influence on human life.

Greek thinkers such as Hipparchus (about 190-120 B.C.) and Ptolemy (about 100-170 A.D.) puzzled out a system in which these seven bodies move around Earth by geometric rules, though complicated ones, since simple circles did not work.

Then, from the scientific revolution of around 1600—specifically from Copernicus's theory in his *De Revolutionibus* of 1543, Kepler's mathematical analysis in his *Astronomia Nova* of 1609, and Galileo's telescopic discoveries reported in his *Sidereus Nuncius* of 1610—it was recognized that five of the moving lights, also our Earth, are globes revolving around the Sun and reflecting its light. So that became the transformed meaning of the word *planet*.

Herschel with oboe and telescope

The first addition to the classic family of six planets came in 1781.

William Herschel (1738-1822) was from a family of Moravian Jews who had moved to Germany and converted to Lutheran Christianity. Like his father, he was an oboist in the military band at Hanover. In 1757, for political reasons, his father sent him to England, then under the Hanoverian kings (the four Georges). He had a prolific musical career; he played not only oboe but violin and organ; composed 24 symphonies. After musical appointments at several cities in northern England, he settled at Bath in 1766.

Through a friend, John Michell, who was an amateur musician and professional scientist, Herschel became interested in astronomy. He started building his own telescopes, of increasing size and quality, which he set up in his back garden. His sister Caroline in 1772 came from Germany to live with him, initially as a singer. She became his astronomical collaborator, and was one of the most prolific discoverers of comets. Herschel made a quarter-century study of double and multiple stars, publishing several catalogues of them and proving that most were gravitationally bound systems. His telescopes were powerful enough to show that many of the comet-like "nebulae" in the Messier catalogue are star clusters, and he published in 1802 the larger Herschel catalogue of deep-sky objects. The variety and beauty of their shapes struck him as a "luxuriant garden."

He made many other contributions to astronomy: pioneered the use of prisms to analyse the spectra of stars and hence their temperature; discovered infrared radiation; discovered that the polar icecaps of Mars vary seasonally.

One of Herschel's telescopes, in the grounds of the Greenwich Observatory.

The discovery of Uranus, 1781

It was during his double-star survey that, on 1781 March 13, Herschel noticed, "in the quartile near ζ Tauri," something that, instead of being point-like, showed a tiny disk; so he thought it might be a star that was nebulous, or large enough to be resolved, or a comet. On later nights it had slightly moved, so he announced it, on April 26, as a comet. Nevil Maskelyne the Astronomer Royal, and others, suspected it might be a planet. From the many positions measured by Herschel and others, Anders Lexell in Russia computed its orbit: nearly circular, unlike the orbits of comets.

(Lexell, a member of the Swedish community in Finland, was one of the intellectuals invited to Russia under the empress Catherine. Comet Lexell (D/1770 L1) was one of Messier's discoveries but named for Lexell because he calculated its amazing path. It passed closer to Earth than any other comet, so close to Jupiter that it went through that planet's satellite system, and disappeared—hence the "D" of its designation—possibly flung out of the solar system.)

The announcement of the seventh "primary planet" made Herschel famous. King George III gave him a stipend of £200 a year—on condition that he move to Windsor as Court Astronomer so that the royal family might look through his telescopes. He received grants for new telescopes; in 1816 he became Sir William Herschel; in 1820 he became the first president of the Royal Astronomical Society.

Uranus pre-discovery

Uranus is visible—just—to the naked eye, easily to the telescope, and had been earlier observed as a "star" at least thirty times. Pierre Charles Le Monnier (whose father was also an astronomer) recorded it at least twelve times between 1750 and Herschel's discovery, including four successive nights, apparently without noticing that it moved. Flamsteed, the first Astronomer Royal, entered it in his 1690 catalogue as 34 Tauri. This was the earliest certain observation, though there is a possibility that Hipparchus recorded it in 128 B.C.

The naming of Uranus

All five old planets bore the names of Roman gods. Herschel thought this unworthy of "the present more philosophical era"; the name should commemorate the date of discovery, and, asked by Maskelyne, he chose Georgium Sidus, "the Georgian Star," in effective flattery of the reigning king.

This was not welcomed outside Britain. French and Swedish astronomers suggested "Herschel" and "Neptune." The proposal of "Uranus" came in 1782 from Johann Bode, now known chiefly for "Bode's Law" describing the spacing of the planets.

Ουρανος was in Greek the sky, or the god of the sky. As Bode argued, Saturn was the father of Jupiter, so the new planet should be the father of Saturn, who was Uranus. The solution isn't perfect: in Greek mythology Uranus is indeed father of Cronus, and Cronus of Zeus; and the Romans equated Cronus with their god Saturn and Zeus with Jupiter; but they had no equivalent of Uranus. So it's a Greek exception in the otherwise Roman family.

The disagreement dragged on. One of the moves in favor of Bode's suggestion was that Martin Klaproth named the atomic element he discovered in 1789 "uranium." The British Nautical Almanac Office kept on using "Georgium Sidus" until 1850.

English-speakers are squeamish about "Uranus" because both of its accepted pronunciations, with stress on the first or second syllable, make them snigger.

Among the nine Muses of the Greeks, Urania was the Muse of astronomy. The large observatory built about 1580 for Tycho Brahe, at the expense of the king of Denmark, on an island between Denmark and Sweden. was named Uranienborg (or in Swedish Uraniborg), "Urania's castle".

Comparative brightness. In the astronomical "magnitude" scale, each step represents a factor of about 2.5 in quantity of light. (The Sun's magnitude is about −27; the Full Moon's about −13.) The moving bodies change in apparent brightness, so for each of them a vertical line represents its range of magnitude. The five anciently-known planets remain above the naked-eye observability limit, except when too near to the Sun. Uranus is just bright enough that it was sometimes noticed as a star before it was recognized as a moving body. Dwarf planet Eris is almost a twin of Pluto in size, but is much more distant. Shown for comparison are selected other objects:
—the brightest satellites of Jupiter (Ganymede), Saturn (Titan), Uranus (Titania), and Neptune (Triton);
—1 Ceres, the asteroid that was first discovered and is large enough to rank also as a dwarf planet, and 4 Vesta, the only asteroid that reaches naked-eye visibility;
—Comet 1P Halley;
—the two brightest stars (Sirius and Canopus), the dimmest of the "top twenty" stars (Regulus), and the nearest star (Proxima Centauri).

The puzzle of Uranus

To calculate the position of a planet, you have to perform a dauntingly long series of computations, of types such as

$$e = 0.0462959 - 0.000027337*t + 0.000000079*t^2 + 0.00000000025*t^3$$

—and then multiply them by the sines and cosines of each other, and a great deal more. Yet these, which computers now do for us in microseconds, were within the capability of mathematicians using pencil and paper. For Uranus they could use not only the recent observations but the pre-discovery ones, covering a longer arc of the orbit.

Alexis Bouvard, in France, published ephemerides—tables of positions—for Jupiter, Saturn, and, in 1821, Uranus. He noticed that, while Jupiter and Saturn were keeping to his predictions, Uranus was not. He surmised that this divergence might be caused by the gravitational pull of yet another planet, yet farther out, which Uranus must have passed about 1821.

He died in 1843. By 1844, Uranus was off its predicted positions by 2 minutes of arc. This "residual," or discrepancy between predicted and observed positions, was an angle one might think tiny—only 2/60 of a degree—but it was unacceptable: it had to have an explanation. Among suggestions were that Uranus had been knocked by a comet; or was being pulled by a satellite that was revolving around it; or that, at such great distance from the Sun, there was a difference in gravitational law. But likeliest, as Bouvard had suggested, was perturbation by an eighth planet.

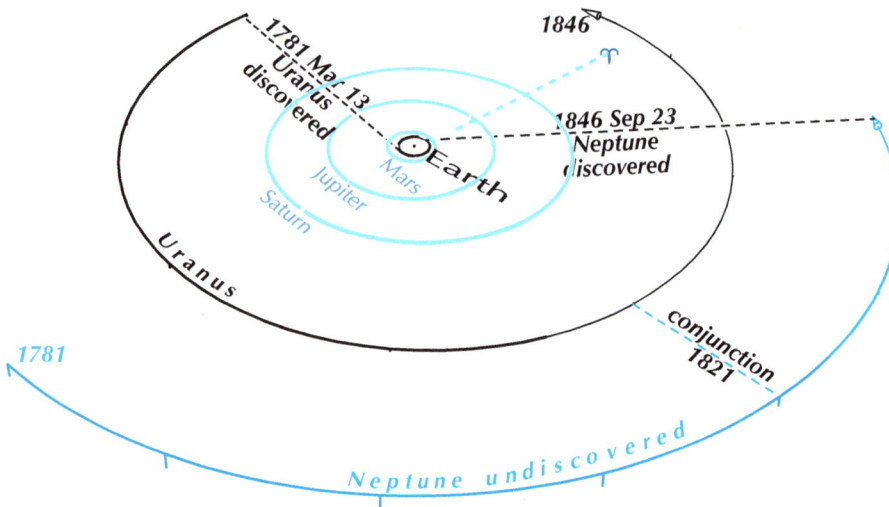

Spatial view showing the paths of Uranus and Neptune from 1781 to 1846. The viewpoint is 40 astronomical units from the Sun and 35° north of the ecliptic plane. The dashed line with Aries (ram's horns) symbol shows the vernal equinox direction.

The quest for a second new planet

"We see it trembling along the far shore of our perception"—a beautiful phrase from Sir John Herschel, William's son, during this time when astronomers were feeling their way out toward this suspected planet, even more remote. (Actually I have yet to rediscover and verify Herschel's wording.)

Some experts thought that centuries would have to elapse before there would be enough data to refine the calculations sufficiently.

But an English student, John Couch Adams (1819-1892), began work on the problem as early as 1841. (His Cornish middle name is pronounced like *cooch*.) In 1843 he became a don at his Cambridge college, and was able to get more data on Uranus's positions from James Challis, the director of the Cambridge observatory, and George Biddell Airy, the Astronomer Royal at Greenwich. Adams worked through the problem several times, coming up with different solutions for the unseen planet's position.

In September 1845 he apparently communicated his final prediction to Challis. On October 21, returning through London from a trip home to Cornwall, he twice tried to visit Airy at Flamsteed House, the Astronomer Royal's residence in the Greenwich observatory. Adams got no answer at the front door. The sequence could make a stage play. But Airy was not "out," as the story is sometimes told and as Adams perhaps thought: the family was not answering knocks at the door, because Airy's wife was suffering a miscarriage.

(She was the beautiful Richarda Smith, whom Airy at age 23 had met while he was on a walking tour in Derbyshire; "Our eyes met and my fate was sealed"; but her father would not let them marry till six years later, when Airy was a professor at Cambridge. They had nine children, of whom the first three died young and the others had various scientific and technical careers.)

Adams left a written note; Airy, finding it insufficiently detailed, sent a question; Adams did not bother to answer, because (according to various later interpretations) he thought the question "trivial," or because he was somewhat feckless. Thus British astronomy missed its chance.

Meanwhile at the Paris Observatory its director, François Arago, suggested to Urbain le Verrier that he study the unknown-planet problem. Le Verrier's name means "the glass-blower," which might have suited a telescope-maker, but he was an expert on the motions of planets and comets. Arago later called him "the man who discovered a planet with the point of his pen." Le Verrier and Adams, unknown to each other, were at work on the problem at the same time, Le Verrier starting and finishing later. He announced his prediction to the French Academy on 1845 November 10. He, too, had ill luck in getting the astronomers of his own country to make a search.

Airy, reading Le Verrier's paper, and remembering about Adams, launched a belated British effort. Challis began a laborious six-week telescopic search on 1846 July 29, a time of year when the Capricornus-Aquarius region begins to be observable in a dark evening sky. He had no success, mainly because he lacked up-to-date charts which would have enabled him to detect the "star" that had moved. It was afterwards realized that he had unknowingly seen the planet, on August 8 and 12.

On 1846 September 18, Le Verrier mailed his prediction to Johann Galle of the Berlin Observatory. Apparently the reason he thought of doing this was that Galle, a year earlier, had sent a copy of his doctoral thesis, hoping for comment from Le Verrier, who hadn't responded, and may have felt he owed Galle a letter.

The discovery of Neptune, 1846

Le Verrier's letter arrived on the morning of 1846 September 23. Galle had a student assistant, Heinrich d'Arrest. (Both of them are remembered as discoverers of three comets each, one being periodic 6P d'Arrest which returns every 6½ years.) D'Arrest pointed out to Galle that they had a recent chart of the relevant starry region. On the evening of the same day, they aimed the observatory's Fraunhofer refracting telescope at that region, Galle looking through it and d'Arrest reading off stars from the chart. Within half an hour they detected the faint "star" that differed from the chart. It was less than a degree from Le Verrier's predicted position, near the border between Capricornus and Aquarius.

It was about a month past opposition, rising not long before sunset. Saturn was only about a degree south of it (the two being in conjunction on March 31 and December 3).

The world found the discovery perhaps even more exciting than that of Uranus. Both resulted from long hard work, but whereas Uranus was a lucky by-product, Neptune was a triumphant validation of scientific theory: of the celestial mechanics proposed by Kepler and explained by Newton.

The credit for Neptune

There was no doubt that Bouvard had been first to suggest the existence of the eighth planet, and that Galle had been first to set eyes knowingly on it. But there was acrimonious debate between Britain and France as to who deserved primary credit for the prediction. After some years, the standard view settled into equal shares for Adams and Le Verrier.

Adams was modest, acknowledging in a paper of 1846 that, though they had worked at the same time, Le Verrier's claim was "just" because he had been the first to publish, and it was his information that had led to the discovery. Now at Cambridge's Pembroke College, Adams did much other painstaking work in mathematical astronomy—the Moon's motion, the Leonid meteors, the Earth's magnetism. But he was so reluctant to compete, publish early, lay claims, stir controversy, or deal with people, that some think he had Asperger's syndrome. When offered the post of Astronomer Royal, he declined it.

There was a kind of two-part verdict: that the British astronomical establishment had lost the race to discover Neptune because of inaction by Airy; but Britain still deserved credit in the person of Adams, whom that establishment had frustrated. Both parts now seem unjust.

Olin Eggen was a prominent American astronomer—he suggested the formation of the Milky Way from a collapsing gas cloud, and the existence of moving groups of stars. At one time (1956 to 1961) he worked at the Greenwich observatory, from which some important historical documents of the 1840s went missing; he always denied having them. After other postings such as in Australia, he spent his last years at the Cerro Tololo international observatory in the high dry Andes of central Chile. He died in 1998, and in 1999 there was found among his papers the missing "Neptune file."

This correspondence showed that Adams was not confident of his own predictions; and that they were too vague, spanning 20° of the zodiac, so it was no wonder that Challis's search failed. Airy's question to Adams was hardly "trivial": he wanted to know what Adams thought the distance of the new planet to be, a point over which Adams had worried.

Mis-statements still occur, as in a quiz question in *The Guardian* (2008 Feb. 9): "Who was the discoverer of Neptune?"—answer, "John Couch Adams."

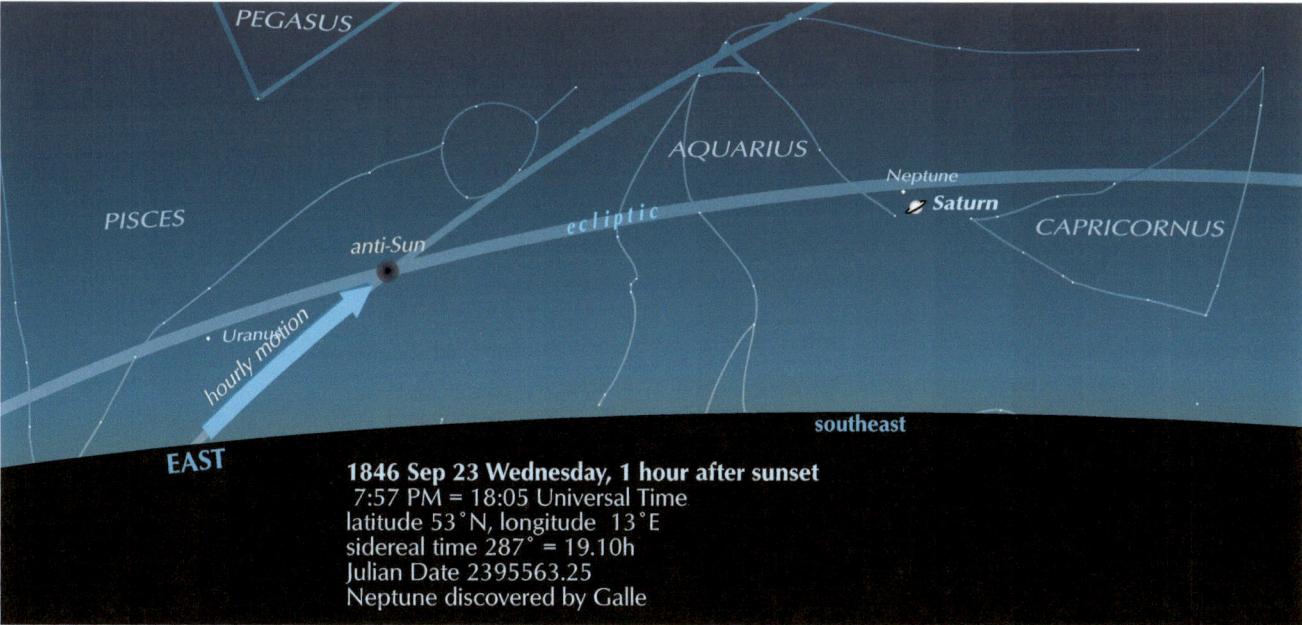

1846 Sep 23 Wednesday, 1 hour after sunset
 7:57 PM = 18:05 Universal Time
latitude 53°N, longitude 13°E
sidereal time 287° = 19.10h
Julian Date 2395563.25
Neptune discovered by Galle

The southeastern sky an hour after sunset, from Berlin on the evening of 1846 September 23. The "anti-Sun" is the direction opposite to the Sun—near to the spring equinox point, since the date is near to the autumn equinox. A broad arrow along the celestial equator shows the angle by which the sky turns in one hour.

Neptune before discovery

Whereas Uranus hovers at the visibility limit for the naked eye, Neptune is about two magnitudes fainter—meaning it sends us 6.3 times less light, and is similar to thousands more stars. Yet, like Uranus, it had been unknowingly observed many times before.

The earliest and most amazing of these pre-discovery observations was by Galileo in 1613. My friend Steve Albers, programmer and weather scientist, had an article in *Sky & Telescope*, March 1979, on very close conjunctions of the planets. He found 21 of them close enough to be occultations—one planet actually getting in front of the other—between 1557 and 2230, with, unfortunately for us, a blank between 1818 and 2065. The illustrations for these rare events were enviable: at most conjunctions, even those we call "close," the planets are dots separated by acres of sky, so it is rich to see two at full size in one frame. One of these rare events that Steve turned up, but which he was not given space to illustrate, was of Jupiter and

the undiscovered Neptune in early 1613. This was soon after the Dutch invention of the telescope in 1608 and Galileo's pioneering use of it on the sky from 1609; in January 1610 he discovered Jupiter's satellites and began making nightly drawings of them. Might he, or others who began to use telescopes, have seen Neptune when it was so close to Jupiter as to appear among the satellites? Steve did suggest this possibility in general terms. The astronomer Charles Kowal and the historian Stillman Drake went to Italy, were allowed to examine Galileo's notebooks, and found he had indeed marked such a "star" in drawings of December 1612 and January 1613, with drawing-details suggesting that he noticed something special about it, presumably that it had moved.

If Galileo had been a bit more specific, he might have to be recognized as discoverer of the eighth planet, 168 years before the discovery of the seventh!

After Neptune's discovery

Le Verrier worked for the rest of his life on sophisticated matters of planetary motion. It was he who discovered that the precession of Mercury's orbit was larger than what it should be, according to Newtonian gravitation, by

about 40 seconds of arc per century. This suggested a search for another unseen planet (Vulcan), on the inner end of the solar system. But the precession was later explained by, and clinched, Einstein's relativity.

14

Pluto's discovery, 1930

The discrepancy between Uranus's predicted and observed positions led to the discovery of Neptune, and that explained the matter—or almost. Figuring Neptune's gravity in reduced the discrepancy to 1/60 of what it had been, mere seconds of arc; yet this, it was thought, still needed explanation. There seemed also to be small discrepancies between the calculated and observed positions of Neptune. So astronomers suspected yet another planet, slightly perturbing the two that had been found.

Percival Lowell, a wealthy Bostonian, founded his Lowell Observatory at Flagstaff, Arizona, in 1894. He believed enthusiastically in the "canals" evidencing civilization of Mars, and in the "Planet X," and searches for it were made from his observatory from 1906 till his death in 1916. William Pickering also in 1909 calculated positions for it, and made unsuccessful searches from the Mount Wilson Observatory.

Vesto Slipher became director of Lowell Observatory in 1926 (his father's surname was Clarke; he apparently took his mother's, and I think I once heard how to pronounce it). In 1929 he re-started the search there, and needed an assistant to do the low-paid work. Clyde Tombaugh, 23-year-old son of a Kansas farmer, had built his own telescopes, and an observatory in the form of an enormous trench that he dug; he sent drawings of Jupiter and Mars to Slipher, who hired him.

Tombaugh's job was to take pairs of photographic plates of identical starry regions a few days apart, and examine them through a blink-comparator. This was a device that by flipping between two such images revealed any "star" that moved. After nearly a year at this task, and many jumping dots that were asteroids, Tombaugh on 1930 Feb. 18 found one that moved along in photos taken on Jan. 21, 23, and 29, and proved to have a distant planetary orbit. It was in Gemini, less than 6° from where Lowell had predicted.

Pluto's name

About a month after the discovery, actually on March 4, Falconer Maddan was reading about it in the London *Times* over breakfast. He was librarian of the Bodleian Library of Oxford University, and his late brother Henry Maddan, a science teacher at Eton school, had in 1878 suggested the names Phobos and Deimos for the satellites of Mars discovered the year before. The newspaper article probably mentioned that names were being sought for the new planet and that the other planets were named for Roman gods, so Maddan's 11-year-old granddaughter, Venetia Burney, said "Why not call it Pluto?"

The family passed the suggestion on, and it eventually defeated others such as Minerva, Zeus, Atlas, Persephone, and Ohnehtn. (Where that last came from I do not know, unless it was a miswriting of the German word *ohnehin.* "anyway.")

A point in favor of *Pluto* was that it begins with Percival Lowell's initials. The symbol chosen for the planet is a monogram of P and L.

Venetia, whose married name was Phair, died in 2009 at age 90. One of the scientific instruments on the 2015 New Horizons spacecraft to Pluto was built and operated by students and was called "Venetia," more fully the Venetia Burney Student Dust Counter.

New Horizons also carried the ashes of Clyde Tombaugh (born 1906 Feb. 4, died 1997 Jan. 17).

Pluto is the Latin spelling of Greek *Ploutôn*, god of death and the underworld. *Ploutos* means "wealth," and according to Plato the god had this name "because from the earth comes up wealth," both vegetable and mineral. The early Romans had gods of the underworld—Dis Pater (Father Dis") and Orcus—but later favored the Greek names, Pluto (usually for the god) and Hades (usually for his realm).

The brothers Zeus, Poseidon, and Hades (Jupiter, Neptune, and Pluto) divided the cosmos between them—sky, sea, and underworld.

Not the predicted planet

Though roughly near to where Percival Lowell had expected Planet X to be, Pluto was not really the planet that had been predicted. It was a coincidence: a fresh discovery by Clyde Tombaugh.

The remaining differences between the calculated and observed positions of Uranus and of Neptune, which had prompted the search, were not real. Pluto was not needed to explain them. They almost disappeared when the planets' distances became better known, and were smaller than the margins of error in the observations. This was proved in 1992 by Myles Standish, using the smaller mass of Neptune found by the Voyager 2 flyby of 1989.

The expected planet would have been larger than Mercury; Lowell expected it to be 6.6 times more massive than Earth. But Pluto turned out to be much smaller than any of the major planets, and it kept being downsized by new research, especially after its satellite Charon was discovered in 1978. This enabled their combined mass to be calculated; and occultations of each by the other revealed their size. Pluto is smaller not only than the eight planets but than at least seven of their satellites. The current estimate of its mass is 0.00218 of the Earth's, and of its width, 0.1868 of the Earth's (0.69 of the Moon's).

Its orbit, much more inclined and eccentric than those of the major planets—and crossing one of them—is more like the orbits of many comets and some asteroids. And it rotates on its side (with axial tilt of 120°), though it shares that oddity with Uranus.

The first of a new family?

Even at the time of Pluto's discovery, some saw that it might be something other than a ninth and smallest planet.

Clyde Tombaugh himself suspected that there might be more Plutos to be found, and continued his search for several years, without success.

Frederick C. Leonard—he who at the age of thirteen, in 1909, founded a national American amateur astronomical society as a sort of rival to the professional one, though he later became a professor himself—wrote in the *Astronomical Society of the Pacific Leaflet* for August 1930: " . . . astronomers have recognized for more than a century that this system is composed successively of the families of the terrestrial planets, the minor planets, and the giant planets. Is it not likely that in Pluto there has come to light the *first* of a *series* of ultra-Neptunian bodies, the remaining members of which still await discovery but which are destined eventually to be detected?"

And Kenneth Edgeworth, an Irish amateur, suggested in papers of 1943 and 1949 the presence of material beyond Neptune.

In 1950 Gerard Kuiper theorized that there had once been a large number of small bodies orbiting in the region beyond Neptune, from which they had fallen inward as comets. Fred Whipple (originator in 1952 of the "dirty snowball" model of comets, and inventor of a subtle way of shielding spacecraft against high-velocity space debris) gave in 1964 a more prescient description of such a "comet ring," as something still existing, and with Pluto as a member of it.

(For much of this interesting historical information I'm indebted to personal communications from Brian Marsden.)

The distant ring is now called the Kuiper Belt. This seems rather unfair, since Whipple described it well and believed in its continued existence, which Kuiper did not. It consists of planetesimals (clumps of matter that could have gone toward the building of planets), and acts as source of short-period comets.

Transneptunian Cubey-One

On 1992 August 30, David Jewitt and Jane Luu, of the Mauna Kea observatory in Hawaii, discovered what seemed just another extremely faint asteroid: 1992 QB1.

That was its provisional designation. The "Q" means that discovery was in the second half of August; it was actually the 27th in that half-month. Asteroids and other minor bodies are nowadays discovered so routinely, many as by-products of surveys for other purposes, that a system for designating them has to be adhered to, even though it is not the most transparent: letters (excluding "I" and "Z") mean half-months, which are subdivided with letters over again, then numbers.

1982 QB1 proved to be no ordinary asteroid. It is in an orbit averaging about 43.7 astronomical units from the Sun (ranging from 40.8 to 46.6); compare Neptune's 30.1. It was the first transneptunian—other than Pluto and Pluto's satellite Charon.

You may see the term written as *trans-Neptunian, trans-Neptunian object,* or *TNO.*

And, among transneptunians, 1992 QB1 is the archetype for the main subdivision, which are therefore called—from QB1— "cubewanos."

There has also been variety in the ways this particular one is referred to. Minor bodies in due course receive numbers; later their provisional designations may be replaced by names. This one got the number 15760. The discoverers wanted to call it Smiley, but there was already an asteroid 1613 Smiley, named in honor of American astronomer Charles Smiley. 1992 QB1, despite its significance, long went without a name, and was commonly called just that, or just QB1. At last in 2018 it became 15760 Albion.

(*Albion* is a supposed ancient name for Britain; writers connected it with Latin *albus*, "white," as in the white cliffs of Dover, but it probably goes back to old Celtic words. In the mythology developed by William Blake, Albion was a figure rather like Adam. Apparently it is Blake's Albion to which the naming committee referred.)

More transneptunians, and larger

The 1992 discovery inspired a systematic photographic search, in a broad band along the ecliptic, for more of these distant bodies. Five were found the next year; the number grew by 2017 to at least 2,300, though many were lost after discovery.

They are defined as having orbits with average distances beyond that of Neptune. But a few hundred, like Pluto, cross inward over Neptune's orbit. Some, the extreme transneptunians, are at enormously greater distances.

If there were thousands that were fainter than Pluto because farther out, then there was a possibility that among them would be some approaching Pluto in size. And they began to be found: In 2000, 20000 Varuna, roughly 900 kilometers wide. In 2001, 28978 Ixion, 800 km. In 2002, 50000 Quaoar, 1110 km. In 2003, 90377 Sedna, 1000 km. In 2004, 90482 Orcus, 917 km, and 136108 Haumea, 1960 km.

Compare the diameters with those of Pluto, 2376 km; Ceres, 965; Charon, 606. Haumea in 2017 became the first body, other than the four giant planets, known to have a ring. Several of the large transneptunians were discoveries of Chad Trujillo, Mike Brown, and David Rabinowitz at Caltech and the Palomar Observatory.

The apple of discord

In July 2005, Brown, Trujillo, and Rabinowitz reported the finding of two relatively bright transneptunians. (They had intended to take more time for observations before making an announcement that could have consequences, but were precipitated into it by the announcement of Haumea.) One of the new finds was 2003 UB313 (meaning that it was on an image from October 2003). (The other later became 136472 Makemake.)

2003 UB313 proved to be near the outer end of an orbit ranging from 38 to 98 astronomical units. It is slightly smaller than Pluto (2326 kilometers wide) but 27 percent more massive. In October 2005 it was found to have, like Pluto, a satellite, whose motion, as with Pluto, enabled the masses to be calculated. UB313 was evidently denser than Pluto, since it proved to be more massive—the most massive transneptunian, so far.

Pluto being still the ninth planet, the new "tenth planet" filled whole newspaper pages. The team had "won the race" to find "the first new planet since Pluto."

The discoverers called their find Xena, which I supposed was because X is the Roman numeral for "ten" and *xena* is Greek for "guest, stranger, foreigner." Fraid not: it was for "the warrior princess in the television series."

(*The Independent* commented in a jokey editorial, 2005 August 1: "But could naming celestial bodies after fictional TV characters create a nasty precedent? Planet Homer anyone? Who could take space exploration seriously again after that?" I stared at this until I realized that the reference must be not to Homer, father of European literature, but to Homer Simpson, cartoon oaf. The journalist might be capable of believing that Pluto was named for the cartoon dog.)

"Xena" was only a working nickname and the new body became 136199 Eris, and its satellite 136199 Eris I with the name Dysnomia. Eris, "strife," was the mischievous spirit who sent the Apple of Discord (inscribed "For the Fairest") rolling among the goddesses, thus starting the train of events that led to the Trojan War. And Dysnomia was one of Eris's baleful daughters, the name meaning "bad law," though some took it to mean "bad naming." These surely were sly digs by the naming committee, since Eris was the apple that provoked a nomenclature war among the astronomical authorities.

Further xenonymous touches (word coined by me to apply to curiosities of naming): the actress who played Xena was Lucy Lawless; and a book by David Weintraub, *Is Pluto a Planet?* (published 2006), was illustrated by Adrienne Outlaw.

Sizes

Betelgeuse
star
904,000,000 km

Sun
star
1,393,400 km

scale 1 cm to 10,000,000 km

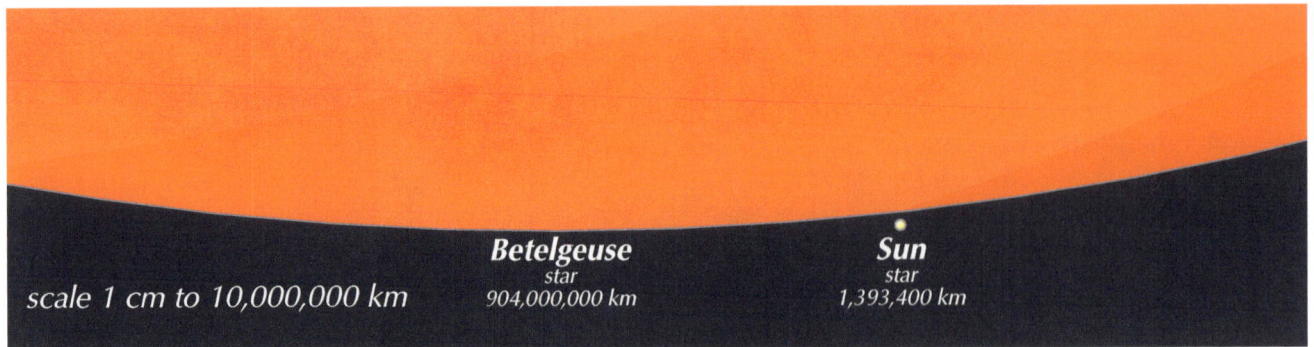

Comparative sizes of a supergiant star (Betelgeuse or α Orionis) and an ordinary main-sequence star, our Sun, with their diameters in kilometers. Betelgeuse is maybe 650 times wider than the Sun, but is difficult to measure and is a variable star: it expands and shrinks irregularly.

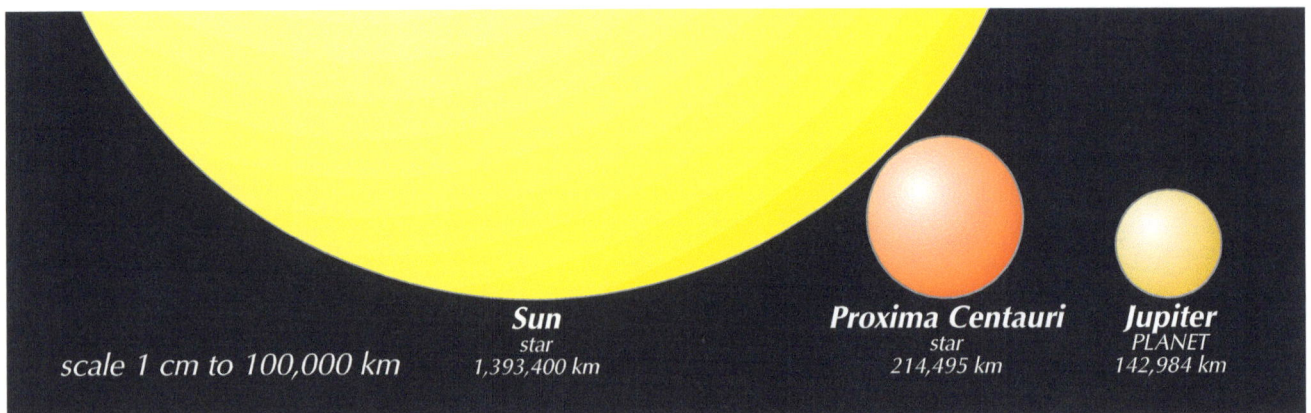

Sun
star
1,393,400 km

Proxima Centauri
star
214,495 km

Jupiter
PLANET
142,984 km

scale 1 cm to 100,000 km

The Sun, a red dwarf star (Proxima Centauri, the next nearest star after the Sun), and the Sun's largest planet.

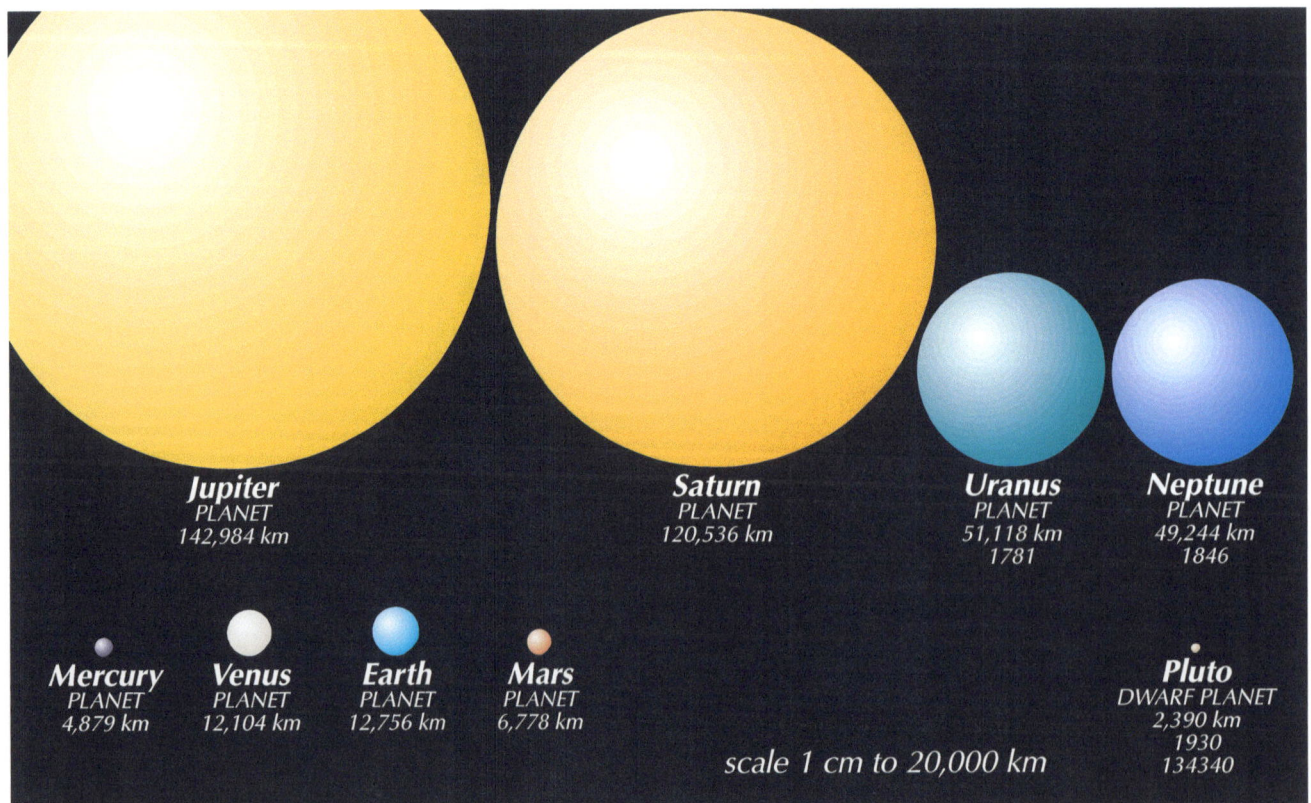

Jupiter
PLANET
142,984 km

Saturn
PLANET
120,536 km

Uranus
PLANET
51,118 km
1781

Neptune
PLANET
49,244 km
1846

Mercury
PLANET
4,879 km

Venus
PLANET
12,104 km

Earth
PLANET
12,756 km

Mars
PLANET
6,778 km

Pluto
DWARF PLANET
2,390 km
1930
134340

scale 1 cm to 20,000 km

The eight major planets, and Pluto. Under their names are their equatorial diameters; and, for the last three, their years of discovery; and, for Pluto, its minor body number.

Ganymede
satellite
5,262 km
1610
Jupiter 3

Titan
satellite
5,150 km
1655
Saturn 6

Mercury
PLANET
4,879 km

Callisto
satellite
4,821 km
1610
Jupiter 4

Io
satellite
3,660 km
1610
Jupiter 1

Moon
satellite
3,475 km

Europa
satellite
3,122 km
1610
Jupiter 2

Triton
satellite
2,706 km
1846
Neptune 1

Pluto
DWARF PLANET
2,390 km
1930
134340

Eris
DWARF PLANET
2,326 km
2005
136199

Haumea
DWARF PLANET
2,322 km
2004
136108

Makemake
DWARF PLANET
1,503 km
2005
116472

Charon
satellite
1,212 km
1978
Pluto 1

Ceres
DWARF PLANET
965 km
1801
1

Pallas
asteroid
550 km
1802
2

Vesta
asteroid
525 km
1807
4

Hygiea
asteroid
431 km
1849
10

Interamnia
asteroid
320 km
1910
704

Europa
asteroid
315 km
1858
52

Cybele
asteroid
300 km
1861
65

Juno
asteroid
300 km
1804
3

scale 1 cm to 2,000 km

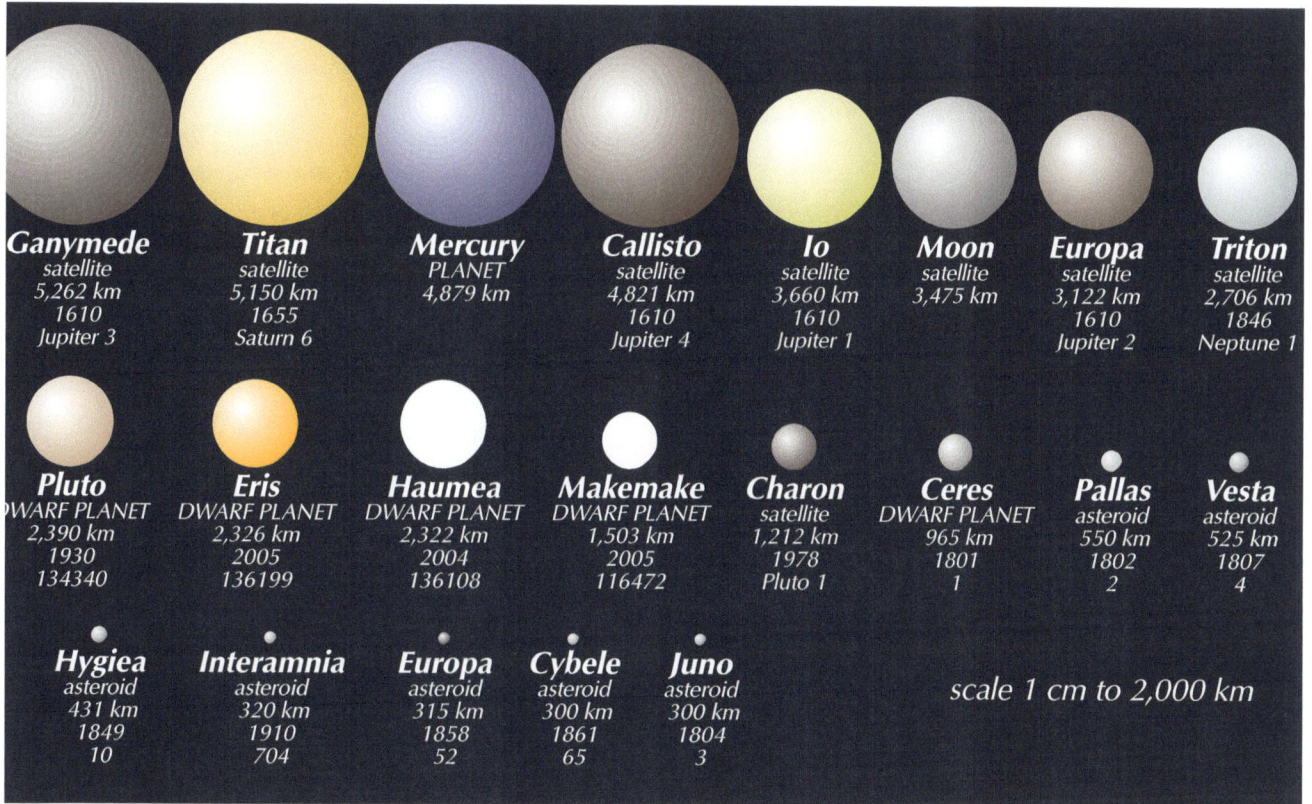

The smallest major planet, and the largest minor bodies, in order of size. Under the equatorial diameter
is given the year of discovery and the minor-body number, or, for satellites, the parent planet.

Moon
satellite
3,475 km

Mercury
PLANET
4,879 km

Io
satellite
3,660 km
1610
Jupiter 1

Europa
satellite
3,122 km
1610
Jupiter 2

Ganymede
satellite
5,262 km
1610
Jupiter 3

Callisto
satellite
4,821 km
1610
Jupiter 4

Titan
satellite
5,150 km
1655
Saturn 6

Ceres
DWARF PLANET
965 km
1801
1

Pallas
asteroid
550 km
1802
2

Juno
asteroid
300 km
1804
3

Vesta
asteroid
525 km
1807
4

Triton
satellite
2,706 km
1846
Neptune 1

Hygiea
asteroid
431 km
1849
10

Europa
asteroid
315 km
1858
52

Cybele
asteroid
300 km
1861
65

Interamnia
asteroid
320 km
1910
704

Pluto
DWARF PLANET
2,390 km
1930
134340

Charon
satellite
1,212 km
1978
Pluto 1

Haumea
DWARF PLANET
2,322 km
2004
136108

Eris
DWARF PLANET
2,326 km
2005
136199

Makemake
DWARF PLANET
1,503 km
2005
116472

Mercury and the largest minor bodies, in order of discovery.

The demotion controversy

It had become obvious that Pluto belongs in the newly-revealed swarm of distant small bodies. It happened to be discovered 62 years before any of the others, because of the historical accident that Clyde Tombaugh was searching hard for an imagined planet, and because of its combination of size and distance. It is at the inner edge of the zone of transneptunians, and is also one of the largest of them. If found nowadays. it would not have been classed as a planet—as was pointed out by Michael Brown, discoverer of Eris, Pluto's rival.

If it was to continue to be a planet, should the same apply to Eris and other large transneptunians? Some at least as large may be found (though they will have to be very dim and very far out). Or should Pluto alone be privileged to remain a planet, on a "historical" if not scientific basis?

A seven-member committee of the International Astronomical Union met in Paris on a June and a July day in 2006, and after anxious discussion came by consensus to this recommendation:

"Planet" should be defined as a body that is massive enough to pull itself into near-spherical shape ("has sufficient mass for its self-gravity to overcome rigid body forces so that it assumes a hydrostatic equilibrium"); is in orbit around a star; and is not a star nor a satellite of another planet. This allowed inclusion of Pluto and Eris, the largest transneptunians (so far); 1 Ceres, far the largest asteroid; and Charon. Charon qualified because, according to this draft, it was more than a mere satellite of Pluto.

Instead of one demotion, three promotions. The number of planets would have risen from nine to twelve. And more, if other large transneptunians were to be found.

But the I.A.U.'s 26th General Assembly, in Prague on August 24, considered this draft and, after angry debate, "came to its senses" (*New York Times*) by adding a further requirement: a "planet" must have cleared out the neighborhood of its orbit—or, as I think I would say, must dominate its orbital region, not sharing it

with others of similar size. This criterion eliminates members of swarms such as the asteroid Main Belt and the Kuiper Belt. Those that did not meet this criterion, but met the others, were to be "dwarf planets."

Those that thus fell from planet to dwarf-planet status were Eris, Pluto, and Ceres. Added were Haumea and Makemake, making five. Four other large asteroids—2 Pallas, 4 Vesta, 10 Hygiea, and 65 Cybele—have been considered for dwarf planet status but rejected. Other large transneptunians are considered by some to qualify; in 2011, Brown listed 390 candidates, some "virtually certain" (Quaoar, Sedna, Salacia, and 2002 MS4); discoveries may raise the number to thousands. A safer term might be "large minor planets."

The revolution in terminology, precipitated by the discoveries of Pluto, 1992 QB1, and Eris, was useful in that otherwise not only these but an indefinite number of others might have to be called planets. The category of "dwarf planets" was introduced as intermediate between "(major) planets" and "minor planets" (which include asteroids and some others), but it is treated as a sub-category of minor planets in that dwarf planets receive designations in the same way and continue to appear among various lists of minor bodies. The major planets returned in 2006 to being eight, as between 1846 and 1930, and will remain so unless another of their size is discovered very far out indeed.

The system of eight major planets in fairly circular orbits returned to being neat. It was very un-neat when little Pluto in its wild orbit was attached to its edge like a sort of ear-ring. Neptune's orbit forms a frontier between the planetary solar system and an outer wilderness.

That is *our* planetary system. It used to be guessed that there were fundamental reasons for it, so that other stars might similarly have a set of inner small and outer large planets in regular orbits. Since 1992, when planets began definitely to be discovered around other stars, we know that other systems can have greatly varied patterns.

Ten thousand

Brian Marsden saw all this coming. British-born, he was director of an institution in Cambridge, Massachusetts, that I thought of as the astronomical clearinghouse, because it had so many aspects. It was the Minor Planet Center, was in the Harvard-Smithsonian Center for Astrophysics, and was the Central Bureau for Astronomical Telegrams, issued for the International Astronomical Union. Brian was the keeper of the records, receiver and publisher of discoveries, and an expert in mathematical astronomy, for instance determining the orbits of new comets and linking some of them to lost comets of the past.

In 1998 he pointed out that the time was approaching when the number 10000 would be assigned to a minor planet. He suggested that the number be reserved for Pluto. Designated as 10000 Pluto, it could be included in lists of similar objects, and would be guaranteed its position at the head of them (the other transneptunians having not yet received num-

bers). Other round-thousand numbers had been assigned in special ways (and have been since: 20000 Varuna, 50000 Quaoar). The giving of such a number would not necessarily imply that Pluto was not a planet.

Those who responded to Marsden's circular voted in favor, but others were indignant. When I asked him where the resistance came from, he said "America. We must have our planet!" Strong objections were coming from the American Astronomical Society's Division for Planetary Sciences. The International Astronomical Union decided against assigning any minor-planet number to Pluto

The result was that, later in 1998, the number 10000 was assigned to a small asteroid discovered in 1951 by A.G. Wilson. (It received the name Myriostos, Greek for "ten thousandth.") And Pluto had to receive, in September 2006, the unremarkable number 134340.

Brian died in 2010.

And yet

Resistance continues. People are sentimental about Pluto, the only "planet" discovered by an American; they knew of it as the ninth planet for 76 years (though few, even of those with telescopes, have seen it). Scientists started a petition; one said "I have nothing but ridicule for this decision." It was pointed out that only 424 of the 2,700 conference attendees were present when the vote was taken; that the existence of Earth-crossing asteroids, and of Trojan asteroids 60° ahead of and behind Jupiter, means that these planets have not "cleared out" their orbits. Bumper-stickers went on sale: "HONK IF PLUTO IS A PLANET."

I feel it takes some perversity to see Pluto as more than a "dwarf planet"—it's rather like calling one of the pedestrians on the sidewalk a car.

Yet Pluto is the only transneptunian you have even a strained hope of observing. At magnitude 14 it is 250 times fainter than Neptune—but as bright as several large satellites, and 40 or more times brighter than the other transneptunians. As a target for the telescopes of skilled amateurs, it continues in the company of Uranus and Neptune.

So I continued to include a Pluto section in my *Astronomical Calendar*. The *Astronomical Almanac*, large volume published by the U.S. Naval Observatory and the U.K. Nautical Almanac Office, which formerly featured Pluto among the planets, changed the title of this to "Planets and Pluto," and subsequently moved Pluto into a new section for "Dwarf Planets."

Spacing of the orbits; the Titius-Bode Law

The speed of the planets' apparent motions gave clues to how far away they were, so it was early understood that for example Saturn was the most remote. Kepler's geometry fixed their relative distances, though these could be turned into absolute distances only when the astronomical unit, the Sun-Earth distance, could be found more accurately from the 17th century onward, particularly from ingenious use of transits of Venus across the Sun.

Was there any regularity to the spacing of the planets? Seventeenth-century habits of mind suggested that there had to be a God-given rule, similar to Kepler's three laws. Kepler himself conceived the beautiful idea that if the planets are set in concentric spheres, then the five regular polyhedra or "Platonic solids" (tetrahedron, cube, octahedron, dodecahedron, icosahedron) could fit between them. He set this forth in his book of 1596, mentioning that the idea occurred to him on July 19, 1595. He also considered, but later discarded, the possibility of an unobserved planet between Mars and Jupiter (and another between Mercury and Venus).

David Gregory in his *Elements of Astronomy* (1715) pointed out that if we divide the Sun-Earth distance into 10 parts, then the relative Sun-distances of the known planets were 4, 7, 10, 15, 52, 96. The distances increase by increasingly large jumps, rather like the widening spiral of a nautilus shell; there was some hint of multiplication by two; but if these were rules they were ragged ones. Charles Bonnet of Geneva remarked in his 1764 *Contemplation de la Nature* that there might be planets yet to be discovered. (He was the first to describe, in 1760, the hallucination syndrome now named for him.) In 1776 Johann Daniel Titius of Wittenberg, translating Bonnet's book from French into German, added to this remark a large footnote in which he suggested a formula to explain the spacing (though he now started by defining the "parts" as 100ths of Saturn's distance). Finally, Johann Elert Bode, earnest life-long astronomer despite his bad eyes, re-stated and popularized Titius's rule, first mentioning it in 1772 in a footnote to the second edition of an elementary textbook he had written, and again starting by dividing the Saturn distance by 100. He later gave credit to Titius, but the rule is more often known as Bode's Law than as the Titius-Bode Law.

If we call the distance of Mercury from the Sun 4, then Venus is at Mercury's distance plus 3, Earth at Mercury's 4 plus 6, Mars at 4 plus 12, and so on.

But there was a gap between Mars and Jupiter, after which the regularity resumed with Saturn. Titius and Bode and others remarked that one of the undiscovered planets suggested by Bonnet might lurk in this gap, which otherwise seemed logically intolerable. And when children walk my *Thousand-Yard Model* of the solar system, their astonishment climaxes at the enormous increase in the paces to be taken after Mars.

Divide the Titius-Bode numbers by 10 and you have the distances in astronomical units.

The formula is not a simple doubling, but incorporates it in a subtle way. It worked pretty well; not precisely, but within a decimal digit or so; well enough to suggest that something valid underlay it.

Then in 1781 came Herschel's discovery of Uranus, at the next outward position in the Titius-Bode series! It fitted, not perfectly but with the approximation characteristic of some other steps. The "law" seemed confirmed, and there was renewed eagerness to complete it by finding the planet in the Mars-Jupiter gap. In 1800, Franz Xavier von Zach sent charts of 24 segments of the ecliptic zone to 24 astronomers of Europe, his "celestial police," asking them to search their segments. They had not even begun when, on 1801 January 1, Giuseppe Piazzi, in Sicily, found Ceres, and its movement showed that it was at the right 2.8 a.u. distance. Ironically, though he was one of those to whom Zach had written, he had not yet received his letter and chart, so he did not know he was supposed to be helping find the missing planet when he happened to find it.

Ceres and the next three similar discoveries—Pallas, Juno, and Vesta, in 1802, 1804, 1807—were until 1846 considered planets. But they were far fainter than the classic six, smaller than the Moon, and proved to be the first of thousands of small bodies around the fifth Titius-Bode position. They were thought to have been a planet that broke up; more likely, one that, gravitationally bullied by Jupiter, never coalesced.

Then Neptune: found in 1846, it fell way short of fulfilling the next outward position. Though the stride from Uranus to Neptune was huge, it should have been 1.76 times huger.

Then Pluto: its average distance, in its very eccentric orbit, is nearer to what should have been the Neptune position. Eris, more massive than Pluto, is far too far out.

Bode's "Law"—unlike the universal laws of Newton that explained Kepler's laws of motion, that in turn explained Copernicus—turned out to be an approximate description that happened to fit much but not all of our star's planetary system.

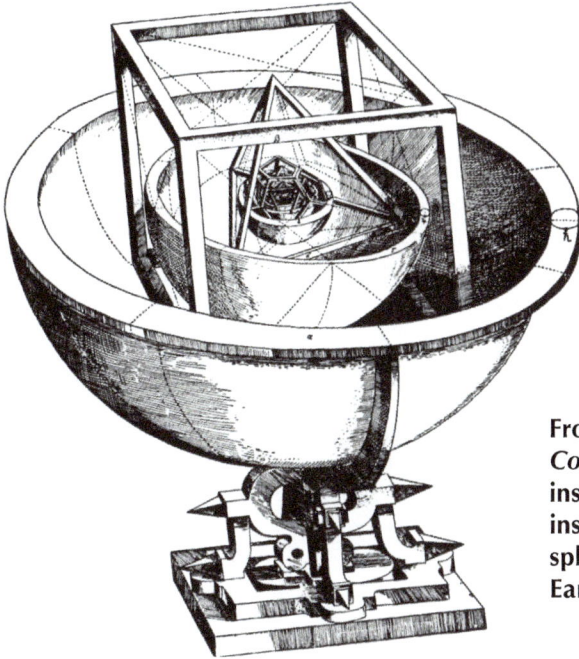

In this table, the third column is the number arrived at by the Titius-Bode formula, and the fourth column is the heliocentric distance (in astronomical units) of an actual body.

4 + 0	=	4	0.39	Mercury
4 + 3	=	7	0.72	Venus
4 + 6	=	10	1.0	Earth
4 + 12	=	16	1.52	Mars
4 + 24	=	28	2.8	(Ceres)
4 + 48	=	52	5.2	Jupiter
4 + 96	=	100	9.5	Saturn
4 +192	=	196	19.2	(Uranus)
4 +384	=	388	30.1	(Neptune)

From Kepler's *Mysterium Cosmographicum*, 1596. Octahedron inside icosahedron inside dodecahedron inside tetrahedron inside cube, separating spheres for the orbits of Mercury, Venus, Earth, Mars, Jupiter, Saturn.

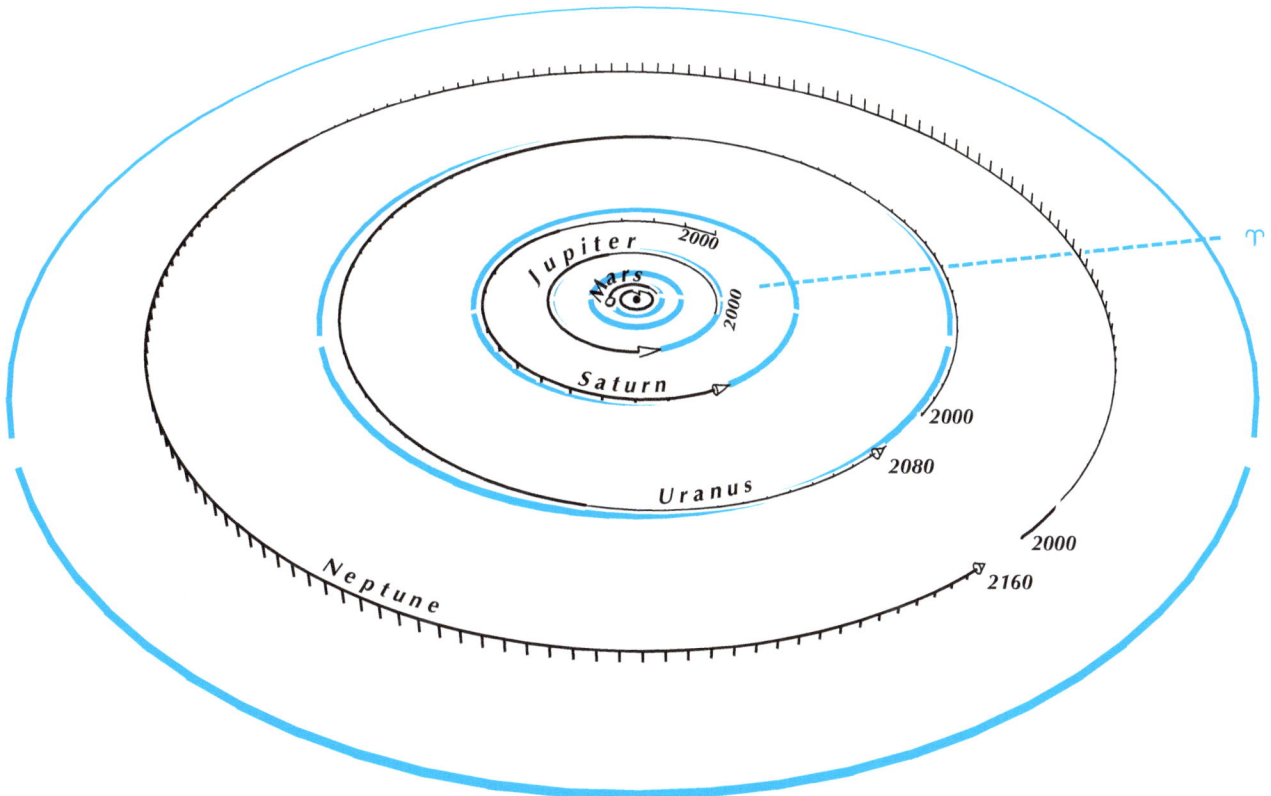

Orbits of Earth, Mars, Jupiter, Saturn, Uranus, and Neptune. (Those of Mercury and Venus are too small to include.) The planets are shown tracing their orbits from 2000 onward. The blue circles are at the distances predicted by the Titius-Bode law. Between Mars and Jupiter a planet is lacking; the asteroids were found instead. Then the fit is fairly good for Jupiter, Saturn, and Uranus; but Neptune disobeys the "law."

Pioneers and Voyagers to the outer giants

Toward the end of the twentieth century, the four giant planets moved into a lucky configuration that I called the Grand Curve: it enabled a Grand Tour, like those that nineteenth-century British aristocrats used to take into Europe. Ahead of Jupiter was Saturn, ahead of Saturn was Uranus, ahead of Uranus was Neptune. Thus a spacecraft by passing close to each could get a gravitational "slingshot" assist that would curve and speed it into a new segment of orbit toward the next. The situation arises rarely, and soon dissolves as the speedier inner planets overtake the outer ones.

Four spacecraft took advantage, and were launched from Cape Canaveral in Florida. Pioneer 10, launched in 1972, became in 1973 the first to fly by Jupiter, and then the first to depart from the solar system. Pioneer 11,

launched 1973, flew by Jupiter in 1974 and Saturn in 1979. Voyager 1, launched 1977, flew by Jupiter in 1979 and Saturn in 1980. Voyager 2, launched 1977 (a few days before Voyager 1), flew by Jupiter in 1979, Saturn in 1981, Uranus in 1986, and Neptune in 1989.

These were the first four solar-system-escapers, the first human-made things to gain escape-velocity from the Sun. "One Planet Awakes" was the title I gave to the diagrams I made of them (from varying viewpoints) to fill pages in several issues of my *Astronomical Calendar*. It was as if the third planet from the Sun had germinated and started sending out spores to the outer system and beyond. They gave us the first close-up views of the four giant planets and their satellites and rings. Voyager 2 is still the only one to have visited Uranus and Neptune.

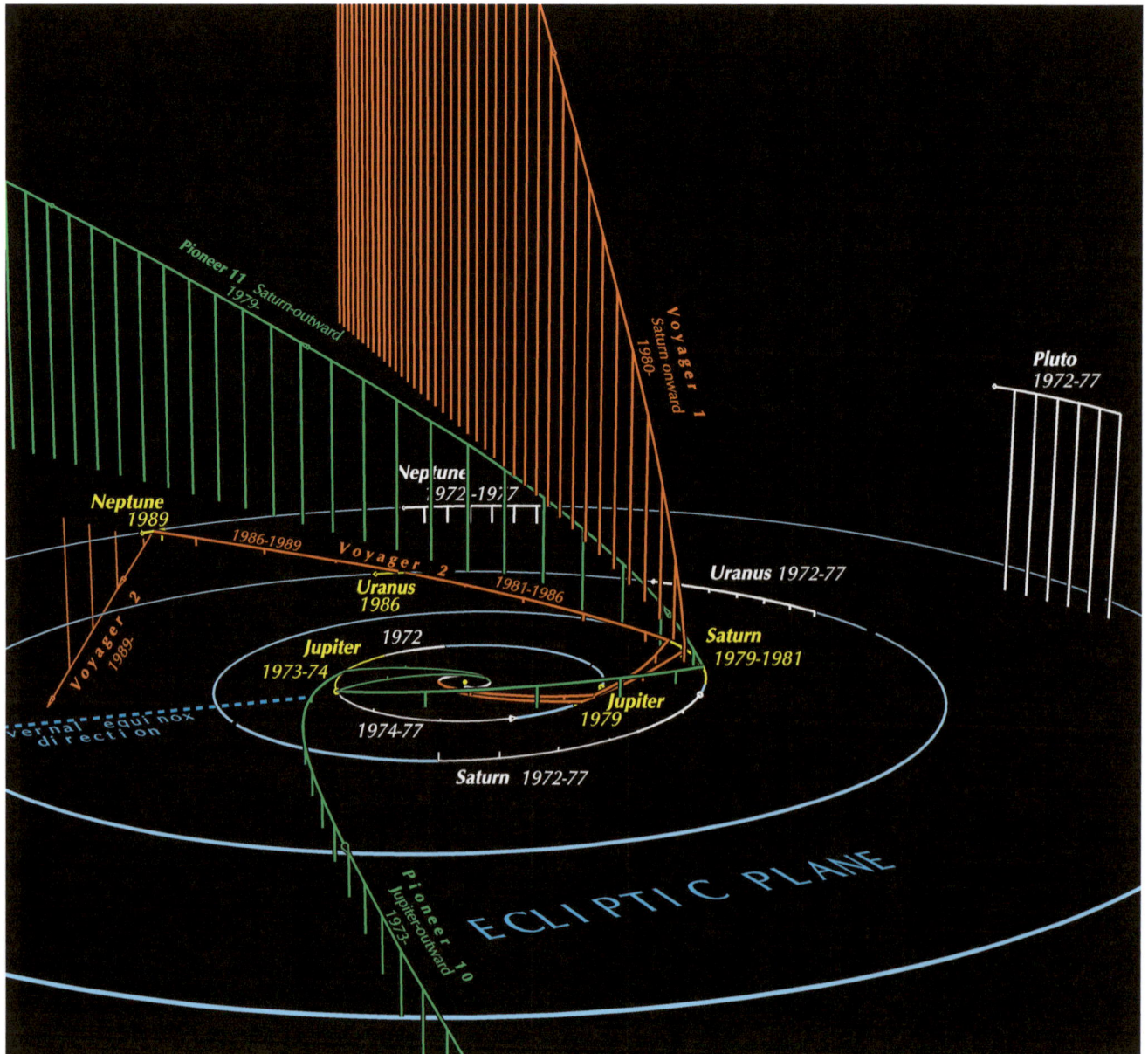

New Horizons to Pluto

Pluto could not be included in the Grand Tour of the outer planets, because it did not fit into the Grand Curve. It was too far back in its orbit; also, it was far north of the ecliptic plane, along which objects can be ejected from Earth with economy of energy. Pluto had to wait for a special mission: New Horizons, planned from about 2000 onward. It was designed for a time when Jupiter would be again in position to give a boost and when Pluto would be almost down to the ecliptic plane.

` The target position was rather close outside where Voyager 2 had met Neptune in 1989, as I realized when I tried to plot all five spacecraft in one picture. It was difficult to find a viewpoint that disentangled their paths; the better solution was two pictures.

New Horizons set out on 2006 Jan. 19 from Cape Canaveral—at record speed. Unlike all previous launches it was already at escape velocity from the solar system. Part of its rocket

launch system is also on an escape path (passing Pluto more widely).

New Horizons flew by an asteroid, which, discovered in 2002, now got the name 132524 APL because the spacecraft had been built at the Applied Physics Laboratory of Johns Hopkins University. And on 2007 Feb. 28 New Horizons passed 2,300,000 kilometers from Jupiter, close enough for a "gravity assist" but not for its course to be much deflected. On 2015 July 14 it flew by Pluto at a distance of 12,500 km and Charon at 28,800.

That October and November, New Horizons had its course adjusted so as to fly by many more transneptunians, the first, on 2019 Jan. 1, having the minor planet number 486958 and the designation 2014 MU69. About 30 kilometers wide, and 44 astronomical units from the Sun, it will be the most distant body visited by a spacecraft and probably the purest example of a remnant from the solar system's formation.

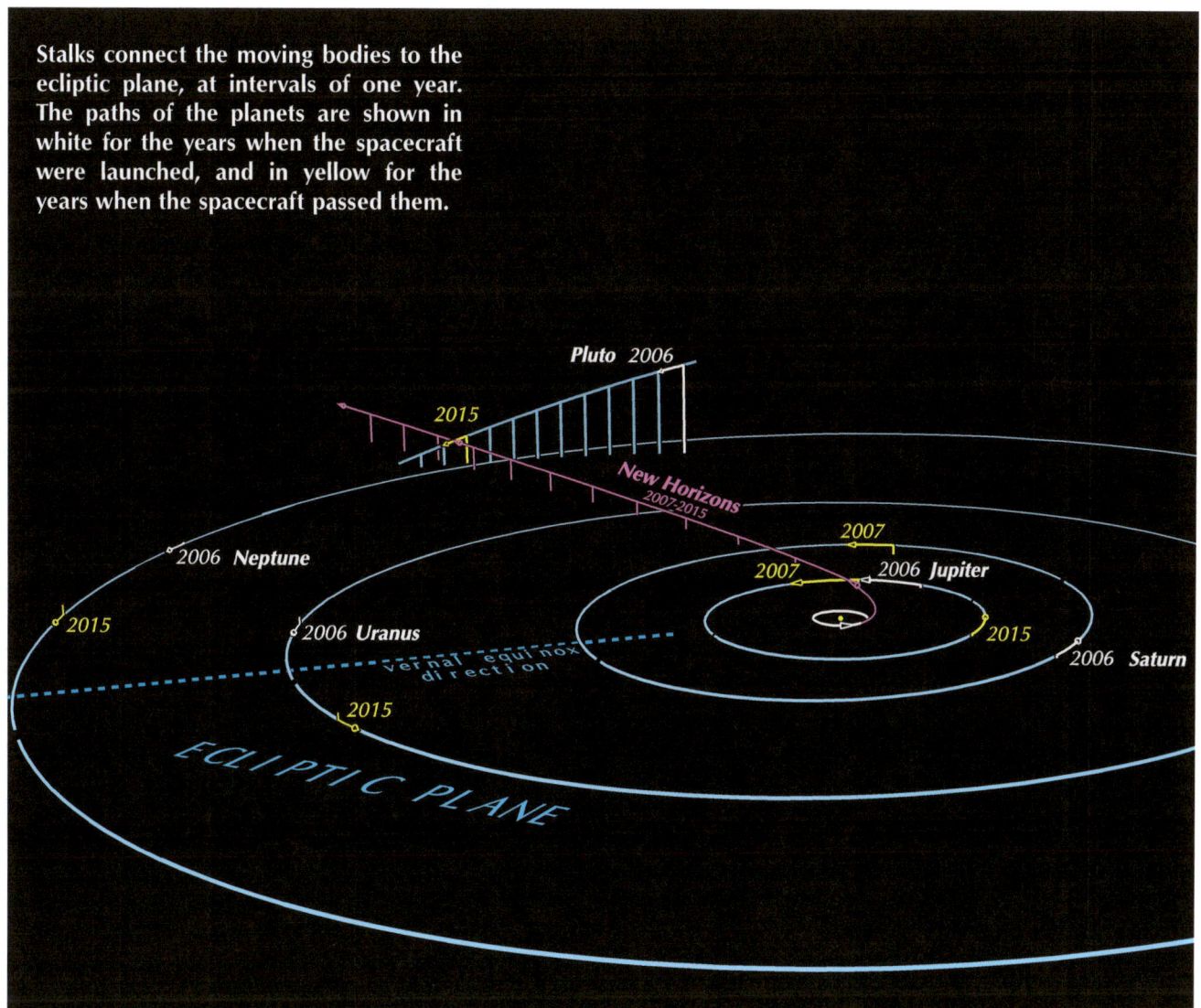

Stalks connect the moving bodies to the ecliptic plane, at intervals of one year. The paths of the planets are shown in white for the years when the spacecraft were launched, and in yellow for the years when the spacecraft passed them.

Missives into the Galaxy

Those five are, so far, the only human-made things that have achieved escape velocity from the Sun and are out on eternal journeys into extra-solar space.

Pioneer 10 is going away toward Taurus—aiming for a point in the Bull's northern horn, about 5° southwest of the horn-tip star El Nath (ζ Tauri). The others are receding in the opposite general direction: Pioneer 11 toward the little constellation Scutum in the dense southern Milky Way, the eastern side of Scutum just south of the ecliptic; Voyager 1 toward northern Ophiuchus, close to its border with Hercules, just south of Algethi (α Herculis); Voyager 2 deep south toward the northern edge of Pavo, just southwest of the Peacock Star, α Pavonis; New Horizons toward northern Sagittarius, near the stars ρ and π Sgr in the asterism called the "Teaspoon."

This does not mean they will reach those destinations. The Peacock Star is about 180 light-years away, El Nath more than 400. The spacecraft are moving at speeds of around

50,000 kilometers an hour—1/190,000,000 of the speed of light. And, a few million years ahead, the stars will be in changed positions. And the spacecraft motions are relative to the Sun, which is advancing around the Galaxy at about 800,000 kilometers an hour. Distant and speedy as they seem to us, the spacecraft are still traveling with the Sun, like a slowly expanding little cloud around it. They are rather analogous to meteors: particles which depart from a comet in all directions but have to follow its general orbit.

Chronology of the Escapers

1972	Mar	3	Pioneer 10 launched
1973	Apr	6	Pioneer 11 launched
1973	Dec	3	Pioneer 10 at Jupiter
1974	Dec	2	Pioneer 11 at Jupiter
1977	Aug	20	Voyager 2 launched
1977	Sep	5	Voyager 1 launched
1979	Mar	5	Voyager 1 at Jupiter
1979	Jul	2	Voyager 2 at Jupiter
1979	Sep	1	Pioneer 11 at Saturn
1980	Nov	12	Voyager 1 at Saturn
1981	Aug	26	Voyager 2 at Saturn
1986	Jan	24	Voyager 2 at Uranus
1989	Aug	25	Voyager 2 at Neptune
2004	Dec	15	Voyager 1 out of solar system
2006	Jan	19	New Horizons launched
2007	Feb	28	New Horizons at Jupiter
2007	Aug	7	Voyager 2 out of solar system
2015	Jul	14	New Horizons at Pluto
2019	Jan	1	New Horizons at minor planet 486958

Neptune, when Voyager 2 reached it in 1989, and Pluto, when New Horizons reached it in 2015, were in nearly the same direction. Since Neptune's mass is about 8,000 times Pluto's, and Voyager 2 and New Horizons passed 4,950 and 12,500 km over the surface of their targets, Voyager 2 was hooked sharply southward, whereas New Horizons was not much deflected.

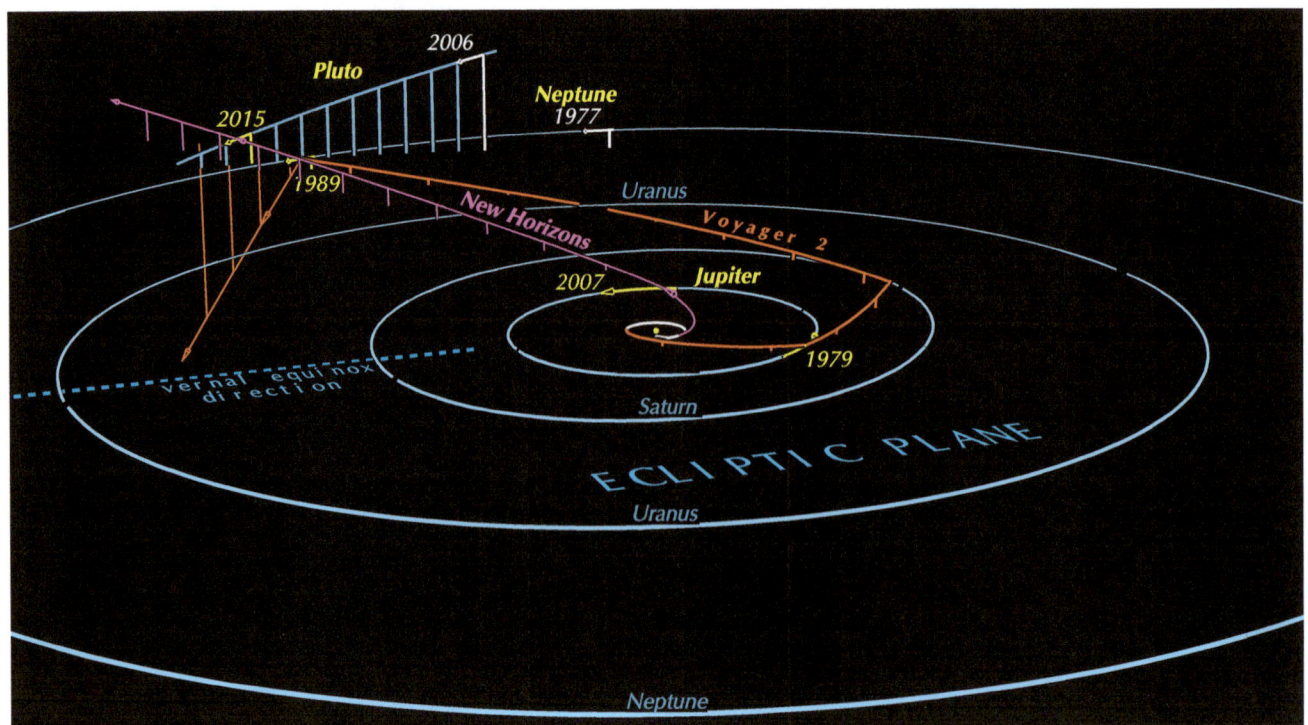

Uranus and Neptune

Uranus's moderate orbit

Uranus travels around the Sun at an average distance of about 19.2 astronomical units. Therefore (by Kepler's third law: period squared equals average distance cubed) it takes slightly over 84 years to move all around the celestial sphere, at an average speed of 6.8 kilometers a second.

So Earth overtakes it on the inside 4.35 days later each year, on average. For instance these oppositions happen at 2016 Oct. 15, 2017 Oct. 19, 2018 Oct. 24, 2019 Oct. 28. So there will come an 84th calendar year when there is no opposition, the previous one being at the end of December and the next at the beginning of January. Such oppositionless years are 1951 and 2035.

The eccentricity of Uranus's orbit is about 0.05, which means that the distance from the Sun varies from about 18.3 a.u. at perihelion to about 20.1 at aphelion. The opposition distance from the Earth will be the same minus about 1— "about," because Earth's orbit, too, is not perfectly circular.

There was for instance an aphelion in 2009 and there will be a perihelion in 2050.

Around the middle of the time between an aphelion and a perihelion, when Uranus is near its average distance, as around 2029, it draws nearer to us from opposition to opposition by as much as 0.069 a.u., which is more than 10 million kilometers. Around the time of aphelion or perihelion, the difference from opposition to opposition approaches zero.

The planet's brightness at opposition varies only from magnitude 5.3 near perihelion to 5.7 near aphelion. A difference of a fifth of a magnitude would be slightly noticeable in two stars side by side, but hardly in Uranuses 42 years apart.

And the width of the planet's bluish disk in the telescope varies from 4.1 seconds near perihelion to 3.7 near aphelion.

The orbit's inclination to the ecliptic plane is 0.77°. Uranus was at ascending node, sloping northward across the ecliptic at this fine angle, in for instance 1945; at greatest northern latitude, by the same small angular span above the ecliptic, in 1965; at descending node in 1984, and at southernmost latitude in 2007.

As seen from Earth, with its own varying distance from the Sun, the relation of Uranus's course to the ecliptic will appear slightly different: for instance when Uranus is near its northernmost latitude, and is at opposition, it will appear slightly farther north from the ecliptic, because Earth is nearer to it than the Sun is.

28

Neptune's circular yet wavy orbit

Neptune, at about 30 a.u. from the Sun is a little over one and a half times farther out than Uranus, so it takes 2.2 times longer to travel—about 165 years.

(More exactly, 164.8. This is a sort of anagram of Neptune's year of discovery, 1846!)

The orbit has the lowest eccentricity of all planets except Venus; in other words, it is nearly circular. So there is a difference of only 0.52 a.u., or about 78,000,000 kilometers, between the perihelion and aphelion distances. This may seem vast, but it is tiny in relation to the 4,500,000,000-kilometer radius of the orbit.

There was an aphelion of Neptune in 1959 and there will be a perihelion in 2042.

However, Jean Meeus in his *Mathematical Astronomy Morsels* (1997) gives several pages (172-178) to showing that what actually happens is curiously complicated. There can be *two* extreme moments around each of such times. Neptune reached a peak of distance from the Sun on 1959 July 13, but then the distance, after decreasing slightly to what Meeus calls a "periheloid" or pseudo-perihelion on 1965 Oct. 6, increased to a second though lower maximum of distance on 1968 Nov. 21.

Similarly, after the perihelion of 2042 September 9, the distance will increase to an "apheloid" in 2049, then decrease very slightly in 2050 to a second and slighter minimum of distance, which is not a true perihelion.

One can see these as oscillations or ripples in the graph of Neptune's distance. They occur all the way along the graph, though only near aphe-

lion and perihelion can they result in double maxima or minima.

Is this because the near-circularity of Neptune's orbit makes it unusually sensitive, perhaps to the gravitational perturbations that all the planets exercise on each other?

No, or not primarily. The cause is that Neptune, like other planets, is really revolving around not the Sun but the barycenter of the solar system. This is close to being the barycenter of the Sun-Jupiter pair (Jupiter having nearly three times more mass than all the other planets together). The barycenter is sometimes inside the Sun, sometimes up to about a Sun-radius outside, depending on whether the planets happen to be all around or mostly on one side. Neptune's orbit around the barycenter makes a smooth ellipse (which reached an "aphelion" at a different time). But, since the Sun and Jupiter revolve around their common barycenter, the distance from the Sun's center to Neptune makes a wavy curve. And when we talk about Neptune's distance, and its perihelion and aphelion, we are referring to the distance from the Sun's center; so it varies in this rhythmic way.

The same is true of all the planets, but with Neptune the effect grows to be noticeable, and sufficient to produce these apparent double extremes, partly because of the near-circularity of Neptune's orbit, but mainly because the Sun-Jupiter distance (5 a.u.) is small compared to the Sun-Neptune distance (30). It's almost as if the Sun and Jupiter are a close binary star with Neptune as a remote third member revolving around them.

Not all of Neptune's perihelia and aphelia have this doubleness, because the ripple in the Sun-centered orbit may coincide too closely with the extreme in the barycenter-centered orbit. (This almost happens in 2042.) But Meeus found that the doubleness happens at *all* the perihelia and aphelia from 1545 to 2300, and then at *none* of them from then onward to 2750. The probable reason is the near coincidence of one orbit of Neptune with 14 of Jupiter, so that similar Sun-Jupiter-Neptune relationships persist for several centuries.

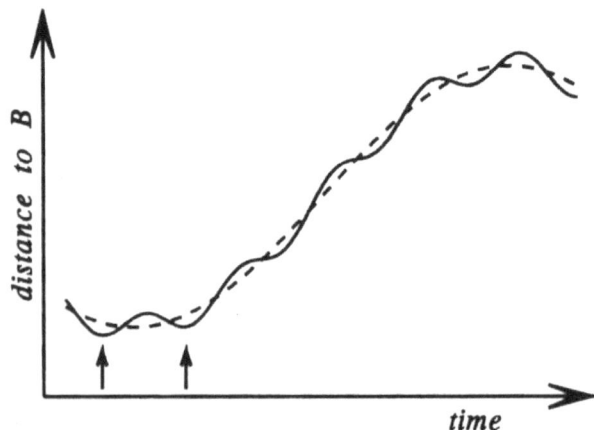

Jean Meeus's schematic graph showing the distance of a secondary body from the primary body of a system (wavy line) and from the barycenter of the system (smoother line, dashed). The arrows show a perihelion and a minimum which is not a perihelion.

Neptune's speed, oppositions, inclination, and nodes

Neptune's average speed in its longer orbit is slower than Uranus's, though not by much: 5.4 kilometers a second. We overtake it at opposition only 2.2 days later each year, for instance 2017 Sep. 5, 2018 Sep. 7, 2019 Sep. 10, 2020 Sep 11 (after a leap day). And it has an oppositionless year once in its 165-year cycle, the next being 2070.

The inclination of Neptune's orbit is slightly greater than Uranus's—by one degree.

It ascended northward through the ecliptic plane at this 1.77° angle in 1920, was at greatest latitude north in 1961, at descending node in 2003, and will be at southernmost latitude in 2044.

Retrograde paths and fivefold conjunctions

While we are overtaking these planets, they appear to move back westward, and since in the middle of these times the planets are nearest to us these retrograde paths can depart most from the ecliptic. But the orbits of Uranus and Neptune have only low inclinations, 0.77° and 1.77°, so their retrograde paths can reach only to these heights as seen from the Sun; slightly more from Earth. In years near the ascending and descending nodes, the retrograde paths are almost straight lines along the ecliptic, closely overlying the direct paths. So it's difficult to distinguish the direct and retrograde parts of the path in a chart!

Uranus is at a distance such that its time of moving retrograde is just 5 months long. And the retrograde paths almost kiss. That is, the point to which Uranus advances one year, before turning back westward, is very close to the rearmost point to which it retreats the next year. (And these points have to be close to the point where Uranus is at conjunction with the Sun, since that comes half way along the direct part of the path.)

Neptune's retrograde paths are a week longer than Uranus's, and they don't just kiss, they overlap. (This even more distant planet is further toward being like a "fixed" star, whose tiny parallactic movement, caused by our shifting viewpoint, stays centered in one place.) The overlapping means that Neptune goes forward, back, forward, back, and finally forward again past any star that happens to be north or south of where its loops overlap. The first two parts of such a quintuple conjunction are in the evening sky, the third behind the Sun, and the remaining two in the morning sky. Such a quintuple conjunction happened for instance as Neptune passed the star μ (Mu) Capricorni in 2009-10, and in 2011-12 with another star even closer north of Neptune's Sun-conjunction point. One could say that it happens with an infinity of faint stars in the background, and with all the stars north and south along the longitude that Neptune is crossing.

The kissing retrograde paths of Uranus, and the overlapping retrograde paths of Neptune.

Geocentric motion

If you look at our charts of the planets' progress, and imagine the ecliptic as the horizon of a frozen sea, Uranus is a skater who skates along a great curve far from us but keeps

executing sweeping loops toward us; and Neptune is doing the same in the more remote distance.

Occultations by the Sun

When an outward planet is traveling close to the ecliptic, that is, it is near to one of the nodes of its sloping orbit, then its conjunctions with the Sun—passages beyond it—can be exact enough that it is actually occulted by the half-degree-wide Sun. These are geometrical events that of course we cannot observe.

Thus Uranus was occulted at its Sun-conjunctions of 1980 to 1989, near to its 1984 descend-

ing node; Neptune from 2000 to 2007, near its 2002 descending node; and Pluto will be occulted on 2019 Jan. 11, near its 2018 descending node. Notice that these occasions are 10 for Uranus, 8 for Neptune, and 1 for Pluto. The very low inclination of Uranus's orbit keeps it lingering near the ecliptic, across which Pluto steeply dives.

How Uranus and Neptune appear to eye and telescope

The disks of Uranus and Neptune are not much different in real diameter: both near to 50,000 kilometers, or about 4 times the width of the Earth. And, despite Neptune's fifty-percent greater distance, they are not very different in angular width from our point of view: Uranus ranges from 3.2 seconds (when it is farthest off behind the Sun) to 4.1 at an opposition near perihelion; Neptune from 2.2 to 2.4. (These apparent sizes are around a thousandth of a degree.)

But Uranus's magnitude at opposition ranges from 5.9 up to 5.2; Neptune's is about 8. In the logarithmic magnitude scale, the brightest stars are around 0, the faintest visible to the naked eye are around 5 or 6, but may be 3 or worse in light-polluted cities, 7 or even better in a clear

and dark sky. So Uranus hovers in and out of this variable naked-eye limit, and this is why it was sometimes seen without being recognized for what it is: plotted as a star (at different positions) many times before it was discovered as a moving body.

The difference of about 2.5 magnitudes between Uranus and Neptune means a difference of about 10 in luminosity—the sunlight reflected to us is 10 times less. So Neptune is difficult to distinguish from the thousands of stars like it or somewhat brighter.

Both planets, seen in good telescopes, are bluish, Uranus verging toward cyan—green-blue—but these tints are subtle. There may be hints of cloud patterns and of darkening toward the limb (visible edge).

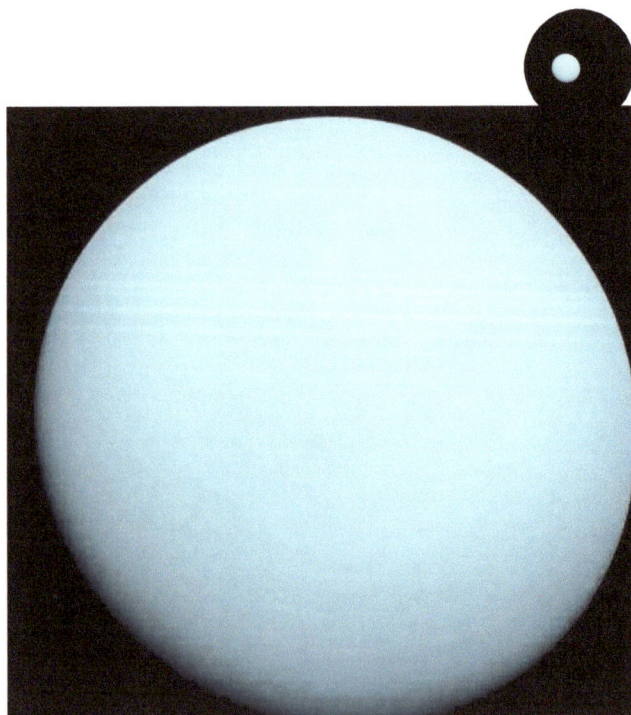

Uranus's more-than-sidewise rotation

All four giant planets rotate surprisingly fast, Jupiter and Saturn in less than half an Earth-day. Uranus is slightly the slowest of the four, taking about 17 hours.

But it is the odd planet in that it rotates "on its side." Instead of spinning roughly in the plane in which it travels—its orbital plane—Uranus is tilted over by about a right angle.

You could picture the normal planet as being like a rubber ball floating on a stream and turning around as it is carried along. Uranus instead is rolling over and over. It rides along with its spin axis pointing horizontally, instead of roughly "up" and "down." For half an orbit one pole points generally forward, until it is pointing outward, after which it is the other pole that is leading around the other half of the orbit.

Uranus's orbital plane differs by only 0.05° from the ecliptic or plane of Earth's orbit; which differs by only 1.65° from the "invariant" plane, or general plane of the solar system, determined mainly by Jupiter.

So which are we to call Uranus's "north" pole? This is a crux, a test case, for the defining of "north" (further discussed in the entry for that word in my glossary book, *Albedo to Zodiac*).

One of the poles points at right ascension 257° (17h 8m), declination –15°, near to the bright star Eta Ophiuchi (Sabik). This is about 8° north of the ecliptic but 15° south of the celestial equator. The direction of the opposite pole is near the top of Orion's bow and near the Taurus border; this is 8° south of the ecliptic but 15° north of the celestial equator. I'll call them the Ophiuchusward and Orionward poles.

The north pole for a planet is defined, by the International Astronomical Union in a 1970 decision, and by others such as NASA and the U.S. Geological Survey, as the rotation axis that lies on the north side of the invariant plane of the solar system, which is little different from the ecliptic plane. By this definition, Uranus's north pole is the Ophiuchusward one, because that is north of the ecliptic plane, even though

it is well down in the southern hemisphere of our sky.

Right-hand rotation

But the planet rotates clockwise as seen from above that pole. So it has to be described as having negative or retrograde rotation—contrary to most of the solar system. Other planets, such as Earth, both spin and revolve counterclockwise as seen from the north (prograde or "right-hand" or "unscrewing" rotation).

Jean Meeus (*More Mathematical Astronomy Morsels*, 2002, 299-301) objects. It would be better to define every "north" pole as the pole from above which the body would be seen to rotate counterclockwise. This definition "does not depend on a particular reference frame, eliminates negative rotation rates, and simplifies the mathematics." I would add that if you switch north and south you switch east and west as well; so on a planet with retrograde rotation the Sun rises in the west—which violates the historic meanings of "east" and "west." And—perhaps the most powerful of Meeus's argument—bodies gradually change their angles of tilt. If that angle happens to be near 90°, as with Uranus, and migrates from one side of 90° to the other, as can happen with smaller bodies such as asteroids, then by the I.A.U. definition we would have to say that the north and south poles have suddenly exchanged places and the direction of rotation has reversed!

This is persuasive, though the difficulty is greater for Venus, which rotates not "on its side" but "upside down": clockwise as seen from its pole which points high into our northern hemisphere. With this as its north pole, it has retrograde rotation. And a vast exception is the Milky Way galaxy: one of its poles is Coma Berenices, 28° into our northern celestial hemisphere, and we really have to call this this the north galactic pole, though the galaxy rotates clockwise around it.

Seasons on Uranus

If the Ophiuchusward pole is the north one, then the axial tilt of Uranus is 98°; if the other, 82°.

Avoiding "north" and "south," we'll say that the Orionward hemisphere (the one Meeus would prefer to call "north") was maximally tilted toward the Sun at Uranus's solstice of 1985 Oct. 6, midsummer for that hemisphere. This was the season witnessed by Voyager 2 in 1986. At 2007 Dec. 16, the planet's equator became edge-on to the Sun; autumn equinox for the Orionward Uranians. At 2030 Apr. 19, again a solstice; midwinter for the Orionwardians. At 2050 Feb. 9, another equinox: spring for them.

Uranus has a brighter polar region on the Orionward side. As this turned away from us around 2007, the planet dimmed for us by about 0.1 magnitude.

As seen from our viewpoint as we go around

our relatively tiny orbit, what happens at Uranus's equinoxes becomes more complicated—a triple event. Though the plane of Uranus's equator—and rings and inner satellites—was edge-on to the Sun at 2007 Dec. 16, Earth, moving out to the "right" (as seen from the north), passed early through that plane, at 2007 May 1. Then our movement took us back across the plane at Aug. 16. Then, out on the "left" side, we crossed for good to the Ophiuchusward side at 2008 Feb. 20.

The seasonal cycle repeats over Uranus's 84-year orbit, though the pattern of the geocentric (as-seen-from-Earth) sub-events differs each time. Thus the "autumn" (for the Orionward hemisphere) equinox has happened 1755 Dec. 1 (before the 1781 discovery), 1839 Dec. 8, 1923 Dec. 12, 2007 Dec. 16.

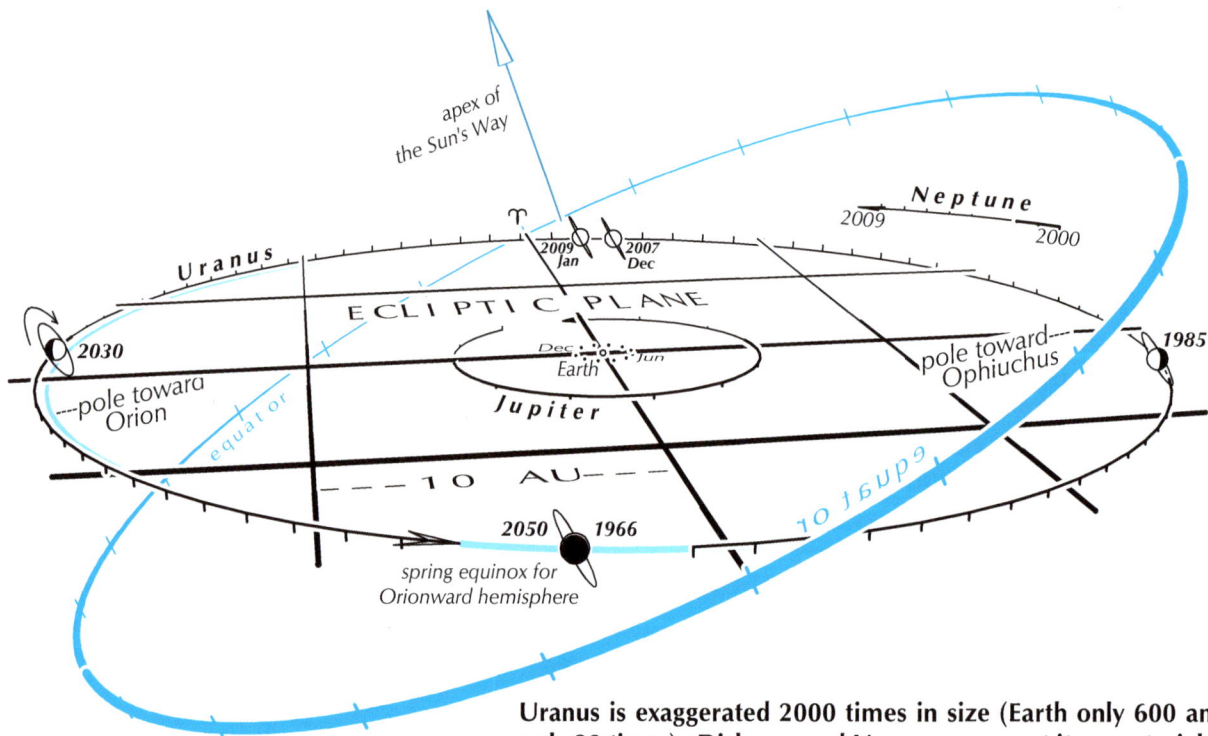

Uranus is exaggerated 2000 times in size (Earth only 600 and Sun only 20 times). Disks around Uranus represent its equatorial plane, or the orbit of one of its satellites or rings. After 2009 they began revolving in apparent clockwise ellipses, to become circles in 2030.

Neptune's rotation

Neptune, like Uranus, rotates rapidly, in about 16 hours. But, unlike the dramatically tilted Uranus, it has a rotation axis at about 28° to its orbital plane; not much different from the 23° and 25° of Earth and Mars.

Its north pole points to a spot at about right ascension 20h, declination 43°, in Lyra, nor far northeast of Vega. And this is not far from what

is called the Apex of the Sun's Way, the direction in which the whole solar system is traveling in its vast orbit around the Milky Way galaxy. The solar system is a disk that happens to stand almost upright as it travels along this part of its orbit, and Neptune's north pole almost points the way.

Clusters of Uranus events

There happen to be approximate groupings of phenomena at the quarter-stages of Uranus's orbit.

The solstices (when the poles point maximally sunward) almost coincide with the ascending and descending nodes of the orbit, so that the equinoxes almost coincide with the times when Uranus is at northernmost and southern-most latitude from the ecliptic.

And these events happen approximately at the cardinal directions of our own orbit. The equinox (spring for the Ophiuchusward hemisphere) happens in the direction of Earth's vernal equinox (spring for our northern hemi-sphere). The Uranus solstices happen in about the same direction as Earth's solstices.

And, thirdly, Uranus's equinoxes happen only a couple of years after its aphelion and perihelion, so its solstices happen at about its average distance from the Sun.

Uranus's nodes and extremes of latitude relate to the orbital plane of Earth, the ecliptic, which Uranus is hardly likely to "feel." More likely, it is influ-enced by the invariant or average plane of the solar system, which differs by 1.65° from the ecliptic.

For instance:

2007 Jan 1: greatest latitude south of the ecliptic
2007 May 1: Earth 1st passage through Uranus's equatorial plane
2007 Aug 16: Earth passage back through Uranus's equatorial plane
2007 Dec 16: equinox (autumn for Uranus's Orionward hemisphere)
2008 Feb 20: Earth 3rd passage through Uranus's equatorial plane
2009 Feb 27: aphelion
2011 Jan 9: arrival at longitude 0°
2011 Apr 9: Uranus crosses celestial equator northward
2011 Oct 16: Uranus crosses celestial equator back southward
2012 Jan 28: Uranus crosses celestial equator definitively northward

Then Uranus proceeds uneventfully around the next quadrant of its orbit, through Pisces, Aries, and Taurus, until there is a complementary cluster of moments around 2030:

2029 May 20: ascending node
2030 Jan 12: at mean distance from Sun
2030 Apr 19: solstice (winter for the Orionward hemisphere)
2032 Dec 23: maximum declination north, 23.65°
2033 Jan 21: at ecliptic longitude 90°

Observationally, the interest is that in 2007-2008 and again in 2030 we see Uranus sideways, with its equator making a north-south line; in the span of years between, we see its Ophiuchusward hemisphere (the northern one according to the I.A.U., southern according to Meeus) gradually opening toward us, to be almost a complete circle around 2018, with its pole in the middle.

Uranus by Moonlight

If a planet's opposition—the middle of the best time to observe it—coincides with the Moon's opposition, that is, Full Moon, then it is near to the Moon and in the middle of its dazzle; whereas if its opposition is timed for New Moon it will not be outcompeted by moonlight. Now, oppositions of Uranus are happening alternately at New and Full Moon. Why?

The average synodic period of Uranus—the time between oppositions—is 369.66 days. The average cycle of the Moon between its "syzygies"—New Moons or Full Moons—is 29.53 days. So the period of Uranus is close to, though not exactly, 12.5 cycles of the Moon. We happen to be in an era when Uranus oppositions are coinciding with New Moons in odd years and Full Moons in even years.

In 2018 the agreement is as close as 2 hours, and, in 2020, closer than 1 hour. The relationship changes gradually, and with irregularities mainly because the Moon's synodic period is quite variable. The syzygies fall increasingly earlier than the oppositions; in 2027 the New Moon will be more than 2 days before opposition.

There will come an era when it is First and Last Quarter that are coinciding with Uranus oppositions.

Is huge remote Uranus somehow sensitive to our satellite? That would be like an elephant trying to keep in step with a fly that is buzzing around you. Or, as some might fancy, there is an occult bond between the Sky God and the Moon Goddess. Not necessary.

The synodic period of a planet is a function of its distance. The more distant, the slower it goes and the sooner we catch up with it; the closer the lap time shrinks toward 365.24 days, the length of our year, which we could consider the "synodic period" of a star—a star comes around to opposition at the same time every year. So there is almost bound to be a planet at a distance where its synodic period is close to 12.5 cycles of the Moon—369.13 days.

Jupiter: 398.88.
Saturn: 378.09.
Uranus: 369.66.
Neptune: 367.49.

In this table the last column shows the difference of syzygy times from opposition times, in fractions of a day.

```
opposition 2017 Oct 19 17h,  New  2017 Oct 19 19h     0.077
opposition 2018 Oct 24  0h,  Full 2018 Oct 24 16h     0.676
opposition 2019 Oct 28  8h,  New  2019 Oct 28  3h    -0.182
opposition 2020 Oct 31 15h,  Full 2020 Oct 31 14h    -0.033
opposition 2021 Nov  4 23h,  New  2021 Nov  4 21h    -0.103
opposition 2022 Nov  9  8h,  Full 2022 Nov  8 11h    -0.883
opposition 2023 Nov 13 17h,  New  2023 Nov 13  9h    -0.320
opposition 2024 Nov 17  2h,  Full 2024 Nov 15 21h    -1.208
opposition 2025 Nov 21 12h,  New  2025 Nov 20  6h    -1.223
opposition 2026 Nov 25 22h,  Full 2026 Nov 24 14h    -1.314
opposition 2027 Nov 30  9h,  New  2027 Nov 28  3h    -2.239
opposition 2028 Dec  3 20h,  Full 2028 Dec  2  1h    -1.773
```

Neptune by Moonlight

Neptune's synodic period is not so close to 12.5 cycles of the Moon—it is 12.44—so it shows something of the same alternation but more rapidly gets out of step.

```
Neptune oppositions, and nearest Moon phases
2017 Sep  5 : Sep  6 FULL
2018 Sep  7 : Sep  9 NEW
2019 Sep 10 : Sep  6 First Quarter
2020 Sep 11 : Sep 10 Last Quarter
2021 Sep 14 : Sep 13 First Quarter
2022 Sep 16 : Sep 17 Last Quarter
2023 Sep 19 : Sep 22 First Quarter
2024 Sep 21 : Sep 18 FULL
2025 Sep 23 : Sep 21 NEW
2026 Sep 26 : Sep 26 FULL
2027 Sep 28 : Sep 30 NEW
2028 Sep 30 : Oct  3 FULL
2029 Oct  2 : Sep 30 Last Quarter
2030 Oct  5 : Oct  4 First Quarter
2031 Oct  7 : Oct  8 Last Quarter
```

Satellites and rings of Uranus

William Herschel, having discovered Uranus in 1781, went on to discover in 1787 the two largest of its five large satellites, Titania and Oberon. (And in 1789 he discovered two of Saturn's: Enceladus and Mimas.)

Not till 1851 were Uranus's next two inward, Umbriel and Ariel, discovered by William Lassell, wealthy Lancashire brewer who built his own observatory and telescopes. Nearly a century passed before the innermost of the five, Miranda, was found in 1948 by Gerard Kuiper. The 1986 flyby of the Voyager spacecraft revealed ten more, and from 1997 to 2003 teams of astronomers have turned up more, bringing the total to 27.

Miranda, Ariel, Oberon, and Titania are characters in Shakespeare's *The Tempest*, and Umbriel in Alexander Pope's satirical epic *The Rape of the Lock*. The tradition of names from Shakespeare and especially his fairy world has continued, with Cordelia, Ophelia, Puck, Prospero, Caliban, Mab.

Satellites I to IV (Ariel, Umbriel, Titania, Oberon) are 1160, 1170, 1580, and 1520 km wide (as against Uranus's 51,100). They shine at about magnitude 14, like Pluto. Titania is the 8th largest satellite in the solar system; smaller than the Moon and the largest satellites of Jupiter, Saturn, and Neptune. The other Uranian satellites are small; 13 are close in, among Uranus's rings, and 9 are in outer and irregular orbits.

Herschel in 1789 thought he saw, and drew a sketch of, a ring around Uranus, but his telescope may not have been capable of it. The rings began really to be discovered in 1977, and 13 are known. They are thin and dark compared with the rings of Saturn, but more definite than those of Jupiter and Neptune.

The chief and inner satellites, and the rings, revolve circularly around the planet's equator, which stands almost perpendicular (at 98°) to the plane of its orbit. This equatorial plane is edge-on to us at Uranus's equinoxes, as in 2007. Around such a time, the satellites appear to move in almost straight north-south lines. For about three years around that time, they and their shadows can transit across Uranus. The first such transit ever observed was that of Ariel on 2006 July 26. After this stage, the apparent orbits gradually opened into ellipses, the satellites being nearer to us when on the east (left), and slightly nearer when at the north than at the south. The apparent orbits will become nearly circular around 2030, when Uranus is at its solstice, its pole toward the Sun.

Satellites and rings of Neptune

Only 17 days after the 1846 discovery of Neptune, William Lassell discovered its largest satellite, Triton. Satellite II, Nereid, was not found till 1949, by Kuiper. Six close-in satellites were discovered when Voyager 2 flew by in 1989, and more later, bringing the total to 14.

The names given to King Neptune's satellites are watery: Triton was a merman-like sea god; Proteus embodied the changeableness of the liquid world; the Nereids and Naiads were nymphs of the salt waters and the fresh; Thalassa simply means "sea"; Psamathe was the selkie who rolled onto the sandy beach and shed her seal skin and became the wife of Aeacus. Sao, I suspected, was the Smithsonian Astrophysical Observatory (stars from its catalog have designations such as SAO 122731), but no, the name does appear in lists of the fifty daughters of Nereus.

Triton is in a circular but highly inclined and retrograde orbit. Nereid is 16 times farther away in a flat but very elliptical orbit. The seven inner satellites are in circular orbits; the five innermost are in synchronous rotation, keeping them stationary above the planet's equator; the five outer satellites, beyond Triton and Nereid, are in orbits with varied eccentricity and inclination, some of them, including Triton, retrograde, that is, revolving counter to the planet's rotation; suggesting that, rather than forming from debris circling the planet, they may have been asteroids of the Kuiper Belt captured by Neptune's gravity.

Triton is 2,700 kilometers in diameter (as against Neptune's 49,500). Proteus is Neptune's second largest; it was found in 1992 to be larger than Nereid.

Triton shines at about magnitude 14, like Uranus's four major satellites and Pluto. The other satellites are much fainter.

Neptune's five rings, discovered in 1984, are so thin that they at first seemed mere partial "arcs." The inner satellites orbit among them. They received grand names from the history of Neptune: Galle, Le Verrier, Lassell, Arago, Adams. There are traces of rings farther out.

Progress of Uranus and Neptune around the sphere

Uranus and Neptune have to creep along their vast orbits relatively slowly; they advance only about 4° and 2° a year.

Uranus since its 1781 discovery has made about 2.8 journeys all around the celestial sphere. It was discovered at longitude 87.5°, the eastern end of Taurus, and returned past that longitude at the end of 1864 and in 1948. By 2017 it was another 298° on, in Pisces.

Neptune since its 1846 discovery has made only a little over one round trip. It returned in 2011 past the place in Capricornus where it

was discovered. July 12 was when it reached the heliocentric longitude (329.3°) where it was discovered; Feb. 11 was the date when, as seen from Earth, Neptune was nearest to its discovery position. This detailed answer was given by Aldo Vitagliano, of the University of Naples, to a questioner in an online eclipse discussion group: "On February 11, 2011, Neptune will appear at the geocentric position closest to that of its discovery, i.e. having the star background closest to that of September 24, 1846, 0:00 UT."

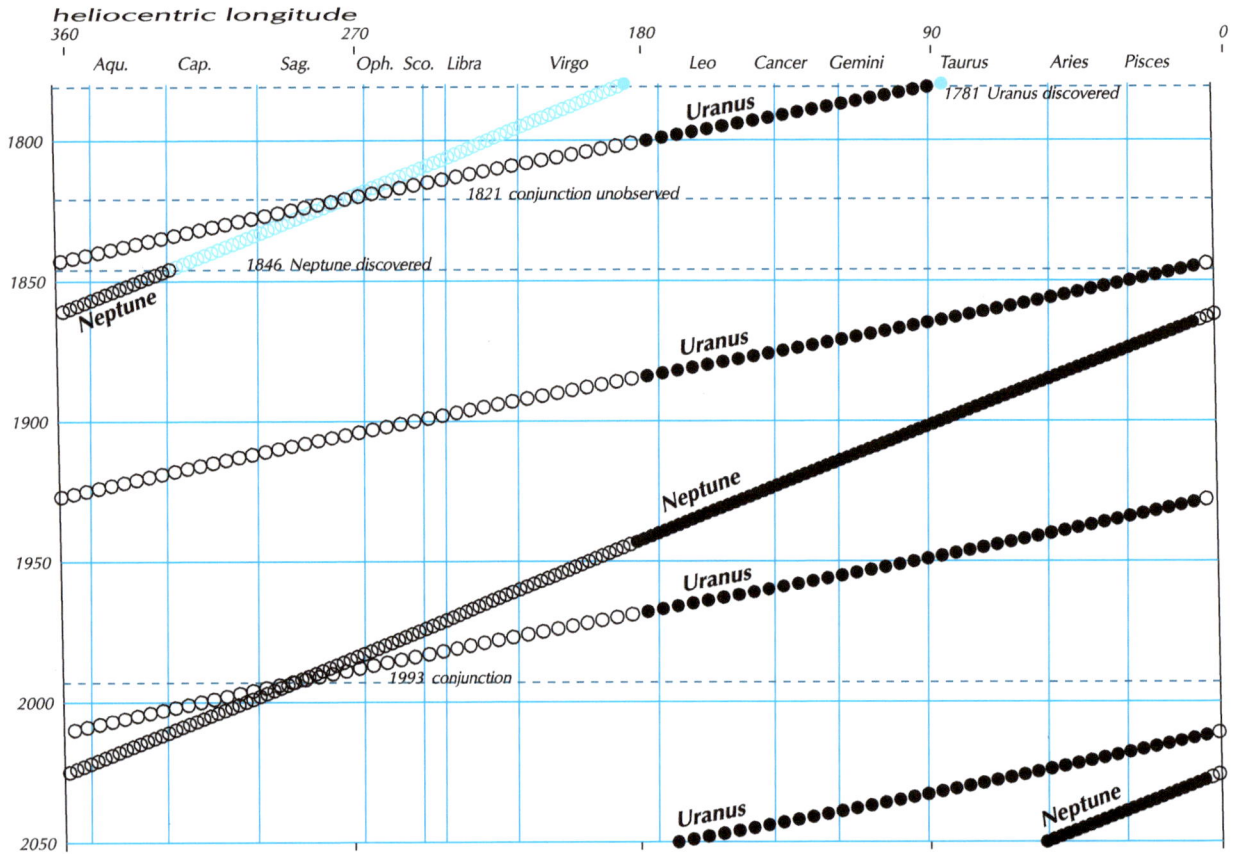

Progress of Uranus and Neptune around the sky since their discovery in 1781 and 1846. They were at conjunction in 1821 (before Neptune was discovered); Uranus then had to travel a little more than twice around the sky to overtake Neptune for the first observed time in 1993. In 2011 Neptune completed its first circuit around the sky since discovery.

The circles representing the planets are blue if before discovery, and solid if in the northern celestial hemisphere.

The lines representing the constellation boundaries slope slightly, because precession causes their longitude to increase slowly (at about 0.014° a year).

Uranus vibrating between north and south

Uranus when discovered in 1781 was in the northern celestial hemisphere, indeed near the northernmost arc of the ecliptic, where Taurus meets Gemini. Moving all around the sphere nearly three times, it has perforce moved from hemisphere to hemisphere six times:

1800-01 to south: Dec 16, Jan 24, Sep 14
1843-44 to north: May 30, Aug 12, Mar 11
1884-85 to south: Nov 14, Feb 27, Aug 29
1927-28 to north: May 1, Sep 16, Feb 21
1968-69 to south: Oct 25, Mar 26, Aug 10
2011-12 to north: Apr 9, Oct 16, Jan 28

Why do we give three dates per event? These are triple events, spreading over about 9½ months. As seen from the Sun, Uranus each time crosses Earth's equatorial plane only once; but, as seen from Earth which circles around the Sun, it seems to cross, then cross back, then cross again. In other words, the path on our sky traced by the planet slants at first forward (eastward) across the equator, then back across it in the retrograde part of the path, then across it for a third and final time.

During the retrograde loop, therefore near to the second part of the triple event, the planet has to be at opposition; for instance on 2011 Sep. 26.

There could be a single crossing of the equator only in the unlikely case that the equator happens to pass through the gap, which as we have seen is small, between the retrograde loops traced by Uranus in two successive years!

Half way between the equator-crossing triple events, Uranus is at its northernmost or southernmost in the sky, at declinations of around plus or minus 23.7°. This is slightly more than the 23.4° northernmost and southernmost points of the ecliptic, because of Uranus's small ventures north and south of the ecliptic.

However, these extremes of declination are not necessarily achieved exactly half way between the equator-crossings. For several years around this time, the planet during its retrograde path moves at nearly its northernmost or southernmost, but which year achieves the actual extreme for the cycle, and on which day, varies according to when opposition happens to fall. For instance, on 2033 May 25, 2034 Mar. 3, and 2035 Mar. 11, Uranus will reach declinations of 23.65°, 23.70°, and 23.63°.

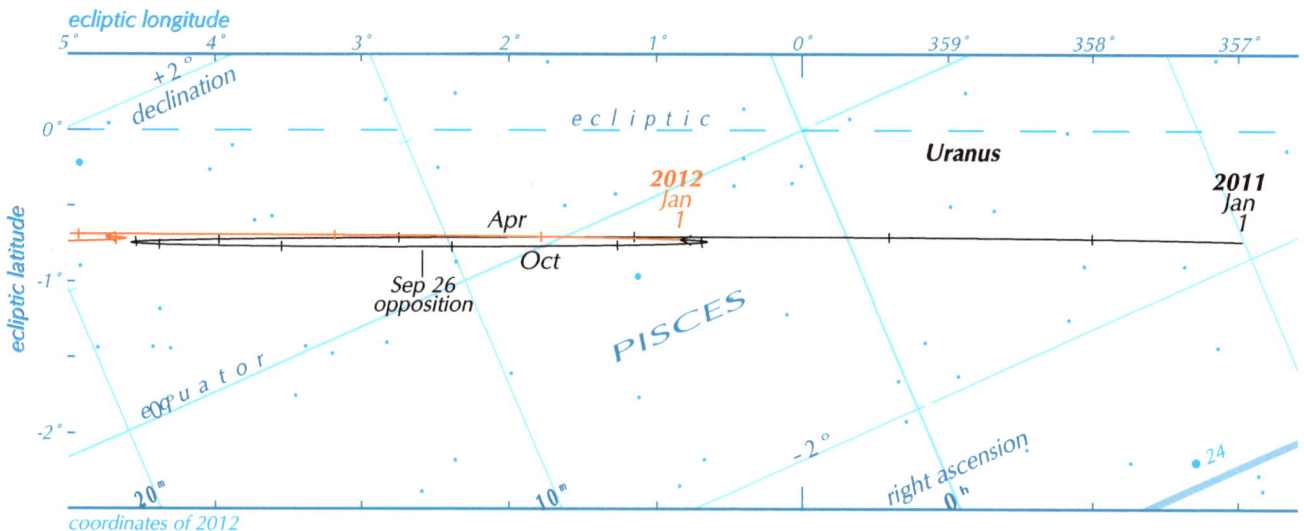

The path of Uranus in 2011 and 2012, crossing the celestial equator three times.

Ecliptic-based chart. The more familiar grid of equatorial coordinates (right ascension and declination) is also shown, sloping and curving in relation to the ecliptic system. For clarity, the second year is plotted with a different color.

Neptune's fivefold waverings across the equator

Neptune happened to be crossing the equator southward in the year (1781) when Uranus was discovered.

Slower-moving and not discovered till 1846, Neptune has since then switched hemispheres only twice, in 1863 and 1944, and will next do so in 2026, rejoining Uranus in the northern hemisphere.

But there is a difference, sometimes, in the way Neptune does it. Because its retrograde paths overlap, it is possible for the celestial equator to intersect both the direct and the retrograde parts of two years' paths. Therefore

the planet advances across the equator a first time, retreats across it, advances across it again, retreats across it again, and advances across it finally. Not just a triple but quintuple phenomenon.

This happened at both the 1821-centered and 1863-centered occasions. In 1779-1781 it caused the total event to spread into parts of *three* years, and in 1862-1863 it nearly did so. But the 1943-centered and 2026-centered occasions are merely triple, with the equator intersecting only two legs of the planet's path.

1781-1781 to south: Nov 24, Feb 7, Sep 20, May 4, Jul 16
1862-1863 to north: May 15, Aug 22, Mar 15, Nov 24, Dec 29
1943-1944 to south: Nov 1, Mar 7, Sep 4
2026-2027 to north: Apr 24, Sep 15, Feb 26

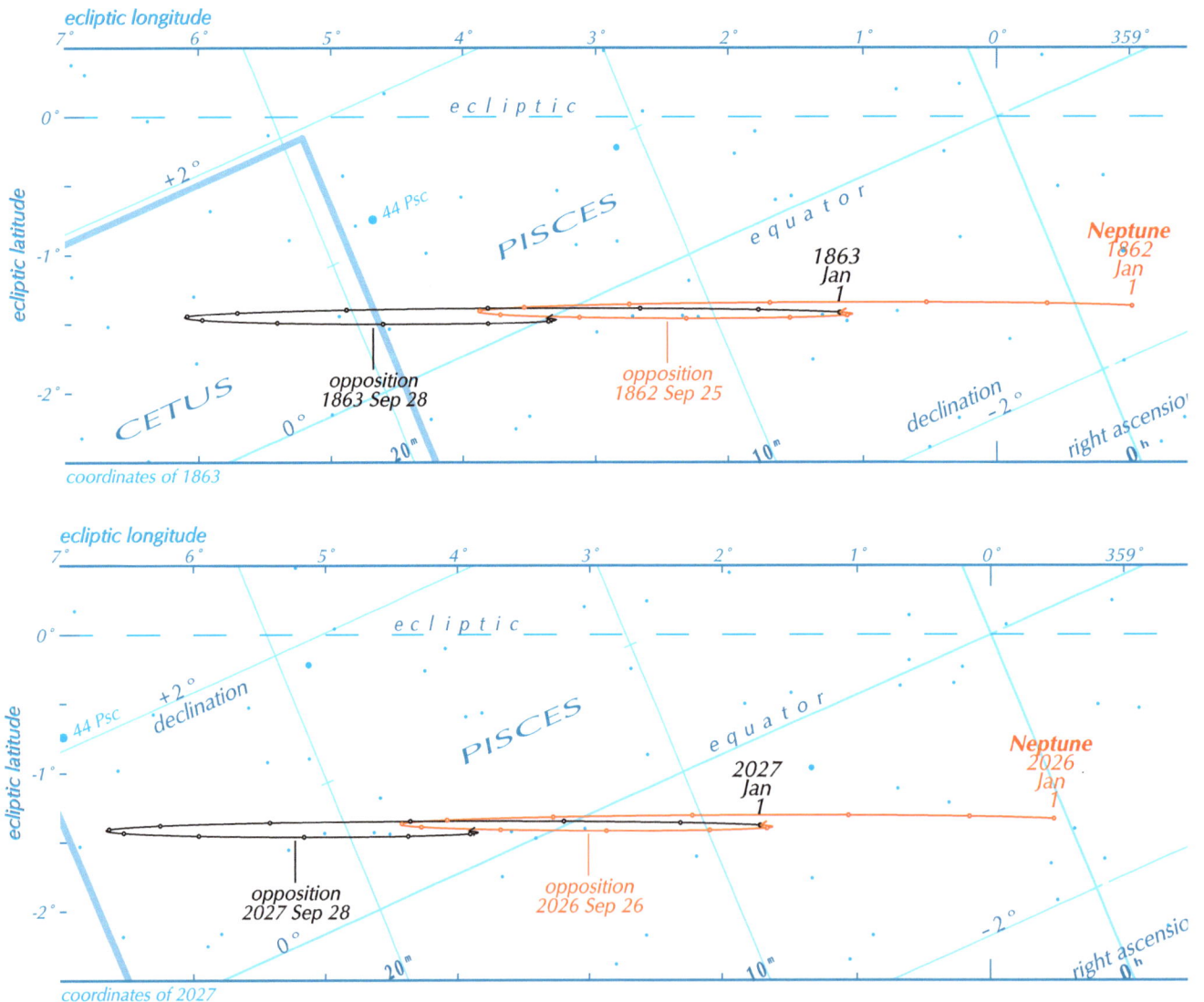

Notice that, after 164 years, the equator-crossing takes place against a slightly shifted starry background. (It has moved farther from the constellation boundary of Cetus and the star 44 Piscium.) This is because precession causes the celestial equator to slide eastward along the ecliptic, at a rate of about 1 degree in 72 years.

Neptune not so far north and south

Roughly half way between its complicated equator-crossings, Neptune is at its northernmost or southernmost. But, because of the orientation of its orbit, with descending node in the Scorpius direction, it is traveling about a degree south of the ecliptic when in the Gemini direction, therefore attains a lesser extreme of declination than any other planet. It reached a northernmost declination of only about 22.4° on 1903 May 18. Similarly, the southern extreme of declination that it touched half an orbit later was only –22.4° on 1985 Nov. 18.

Conjunctions with other planets

Since the remote planets are below naked-eye brightness, while the nearer ones are usually brighter than most stars, conjunctions between them have to be very unequal.

As seen from the Sun, Saturn passes Neptune in 2025 and Uranus in 2032. Since Saturn's own movement is nearly as slow as those of the outer two, it lingers near them, allowing several back-and-forth conjunctions as seen from Earth.

Faster-moving Jupiter overtakes Neptune each 12.7 years (2009, 2022, 2035, 2047) and Uranus each 13.7 years (2010, 2024, 2037).

At these conjunctions, the difference in latitude between the planets is usually more than a quarter of a degree—wide compared with the few-seconds sizes of the planets. But on 2022 June 6 Jupiter will pass only 0.07° south of Neptune.

Mars, in its orbit of 1.88 years or 687 days, passes Uranus at intervals of from 699 to 708 days, and Neptune at intervals of from 693 to 696 days, the background planets having crept slightly forward. The variation is because of Mars's eccentric orbit.

Venus and Mercury, in their 225- and 88-day periods around the Sun, pass both Uranus and Neptune each time, at intervals only a fraction of a day longer than those periods. Venus is never more than about 47° from the Sun, and Mercury never more than about 27°, so all the conjunctions with Mercury and most of those with Venus have to be in twilit sky.

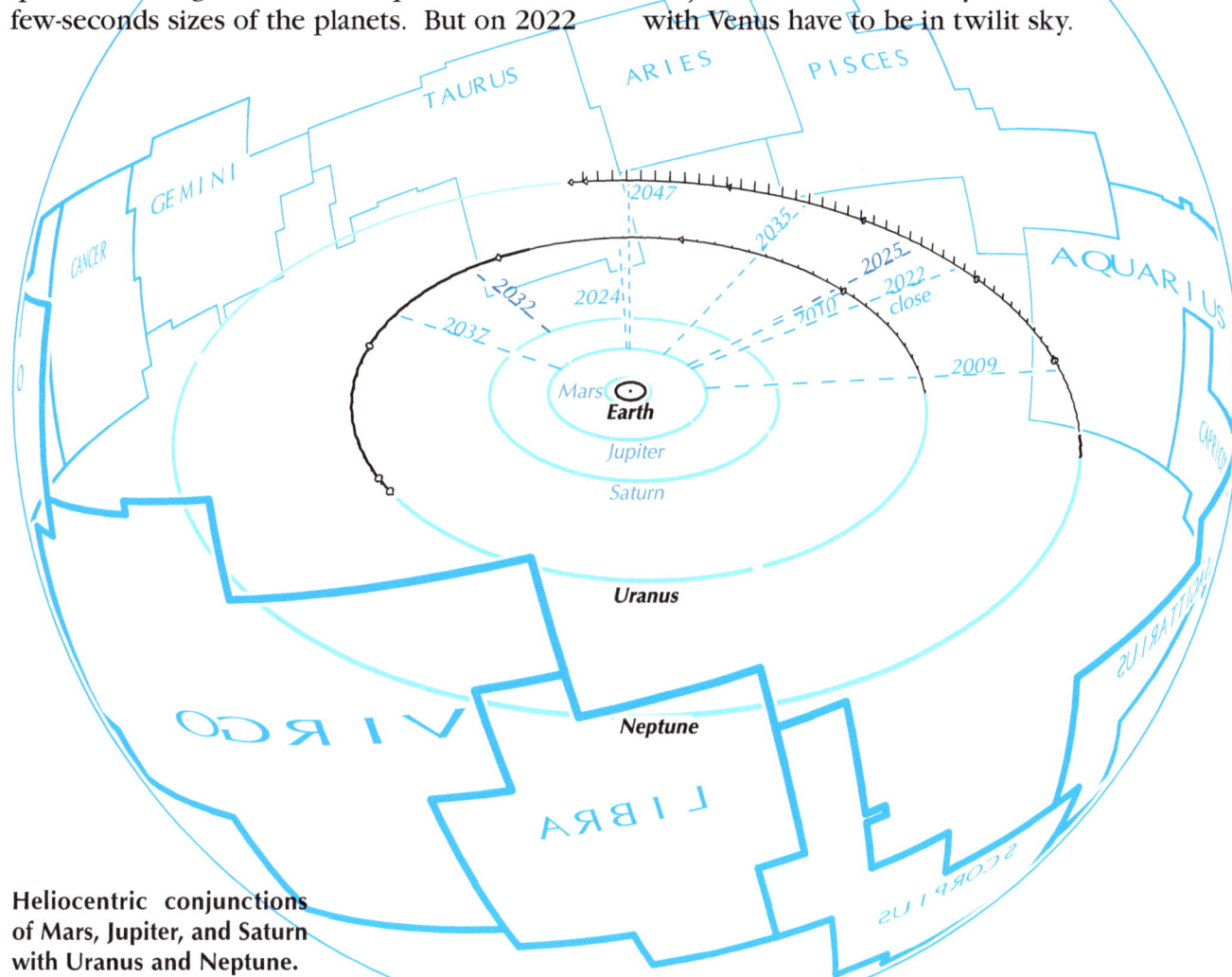

Heliocentric conjunctions of Mars, Jupiter, and Saturn with Uranus and Neptune.

Uranus-Neptune conjunctions

Uranus takes nearly 172 years to lap Neptune. Three of their conjunctions in longitude, as seen from the Sun, are 1821 Sep. 28, 1993 Apr. 20, and 2164 Dec. 17. Each time, the conjunction happens about 17° on eastward along the ecliptic: Neptune has progressed all the way around the sky plus 17°, Uranus twice around plus 17°. So the first of these conjunctions was in western Sagittarius, the second in eastern Sagittarius, and the third will be in Ophiuchus.

Each time, because of the orientation of their slightly inclined orbits, Uranus passes slightly closer to the south of Neptune—1.34°, then 1.12°, then 0.82°. It looks as if there may be an occultation by Uranus of Neptune some centuries ahead, though one has to remember that Uranus appears only 1/900 of a degree wide.

After their unseen conjunction of 1821, Uranus pulled all around the sky, joined Neptune in the southern celestial hemisphere in 1968-69, and overtook it in their first observed conjunction in 1993.

Since then it has been pulling ahead again, increasing its lead by about 1.77° a year. At the 2011-2012 equator-crossing, it left Neptune behind again in the southern hemisphere.

The Uranus-Neptune conjunction of 1821 (before Neptune's discovery) was a normal triple conjunction, as seen from the revolving Earth: the faster planet passed the slower, retreated, passed a third and last time. But the 1993 event, the only one so far observed, was, as pointed out by Jean Meeus in his *Astronomical Tables* (1995), p. 46, a case so rare that it was perhaps unique.

The heliocentric (as seen from the Sun) conjunction of the two planets was on 1993 Apr. 20.

As seen from Earth, and as measured in longitude (parallel to the ecliptic), Uranus overtook Neptune on Feb. 1. Both of them reversed direction in late April. Those events were in our morning sky. The planets were at opposition on July 12, a few hours apart, moving back into the evening sky. Uranus, retrograding faster, moved back past Neptune on Aug. 22. They started forward again at the end of

September; and Uranus finally took the lead (3rd conjunction in longitude) on Nov. 1.

As measured in *right ascension* (parallel to the equator or to a line of declination, such as that at −22°), Uranus first passed Neptune on Jan. 25. They reversed in April. BUT, on Sep. 22, Uranus had retreated ALMOST as far as Neptune before starting forward, without quite consummating the conjunction!

To quote Meeus: "There was no actual conjunction (in right ascension) between the *centers* of the two planetary disks. But as the difference between the right ascension reached a minimum which was almost zero, there was a conjunction between a *part* of the two tiny planetary disks—though in declination the two planets were separated by 1°08′."

This is fantastic. If the planets had been at the same declination—in the same plane—instead of about 50 Uranus-widths apart, the disk of Uranus would have been seen moving with excruciating slowness west, till it covered much but not all of the half-as-large disk of Neptune; hovering there, and slowly reversing.

It was close to a limiting case. The limiting case, essentially impossible, would be for one number (Uranus's right ascension) to shrink *exactly* to the other (Neptune's right ascension) and no further.

In a plan view of these motions, the line joining Earth-Uranus-Neptune on Sep. 22 is *tangent* to the circle of the Earth's orbit. Really, it was the motion of Earth, happening to be at this particular point and curving away around the circle of its orbit, that caused the conjunction of the far-away planets to not-quite-happen.

The next conjunction, in 2164-5, will be triple in *right ascension* but *single in longitude!*

Spatial view of three conjunctions of Uranus and Neptune. The Sun-Uranus-Neptune line shows the heliocentric conjunction of the planets at 1993 Apr. 20. The line from Earth to Uranus's center at 1993 Sep. 22 would pass at that instant between Neptune's center and its forward edge. The ram's-horns symbol shows the vernal equinox direction.

Chart of the 1993 passage of Uranus past Neptune. Their positions are connected by lines perpendicular to the ecliptic at dates of conjunction in longitude (Feb. 1, Aug. 22, Nov. 1) and by thicker lines at the dates of conjunction in right ascension (Jan. 25) and not-quite-conjunction in right ascension (Sep. 22).

42

The ice giants

The four giant planets—Jupiter, Saturn, Uranus, Neptune—were formerly also called "gas giants"; but later it was decided that, strictly, this applies only to the two larger ones. Uranus and Neptune are now "ice giants."

Uranus is slightly larger than Neptune: their equatorial diameters are about 4 and 3.9 times greater than Earth's. (Compare Jupiter and Saturn: 11.2 and 9.45 Earth-widths.) But by mass it's the other way around: Uranus has 14.5 and Neptune 17 Earth-masses. (Compare Jupiter and Saturn: 318 and 95.)

Thus, Neptune, the third most massive planet, is the densest of the four giants. (The four small inner planets are all denser.) By contrast, Uranus, the fourth most massive, is the least dense of all planets except Saturn.

The bodies of Uranus and Neptune can be divided into atmosphere, mantle, and core. But these layers grade into each other; there is no discrete surface. By radius, the atmosphere is roughly the upper 20 percent; the core is the inner 20 percent or less; and the mantle is the bulk, in both radius and mass.

The outer layers, or atmospheres, of all four giants are mainly gaseous hydrogen and helium, with traces of other molecules. One of these, methane, is a cause of the blue coloration of Uranus and Neptune. The atmospheres of both are extremely cold.

The atmospheres are the parts visible to us, so there was much about the two distant planets that could not be found out by remote sensing and had to wait for Voyager 2's passing of them in 1986 and 1989, the launch of the Hubble Space Telescope in 1990, and more powerful ground-based telescopes with adaptive optics to compensate for the unsteady "seeing" through Earth's atmosphere. These revealed dynamic weather on Uranus and even more so on Neptune.

The atmospheres do not have a definite upper boundary; they continue, thinning, to far outside the planets' apparent surface.

Uranus's appearance is bland in visible light, without the cloud bands and storms seen on other giant planets. Winds can carry its clouds along at up to 900 kilometers per hour. But in 2005 there was discovered "the brightest cloud feature ever observed [on Uranus] at near-infrared wavelengths."

By contrast, Neptune's surface shows much activity, weather patterns, differences of color.

It has the strongest winds of any planet, to more than 2,000 km per hour! It had, at the time of Voyager's visit, a Great Dark Spot, a rotating storm like Jupiter's Great Red Spot. Around the hemisphere now tipped sunward are three white cloud-bands, probably made of methane crystals; they brightened greatly after the 1989 flyby.

Below the atmospheres, the mantles are hot dense fluids containing atoms and compounds heavier than hydrogen and helium, such as oxygen, nitrogen, carbon, water, methane, and ammonia. These are referred to by the scientists as "ices," though they are not in a state like frozen water. (In the composition of stars, the heavier-than-helium elements are called "metals." So planets can have lower or higher "iciness" and stars lower or higher "metallicity." One can't help feeling that astrophysicists might have found some term usable in both contexts and less discordant with ordinary usage, but it's too late now.) It is because these ices, which are mainly in the mantle, make up the largest proportion of the total mass that the planets are called "ice giants."

One speculation about the interior of Uranus is that carbon, from the break-down of methane, crystallizes into diamonds, which rain down through the mantle and settle as a diamond ocean.

The cores of Uranus and Neptune are rocky: dense and hot mixtures of iron, nickel, and silicate compounds.

What makes Uranus different from Neptune, and from all other major planets, is its axial tilt of 98° (by the official or Ophiuchusward definition of its north pole) or 82° (by the direct-rotation definition).

How did Uranus get tilted so extremely away from the plane in which most of the solar system's material revolves? It's not known; the usual suggestion is that it collided with a body, of about Earth's size.

The attitude affects climate. The north and south poles of other planets are always near their limbs (edges) as seen from the Sun, and so are cold each night and slightly warmed each day; but each of Uranus's poles spends many years continuously facing the Sun, and then many years facing deep space.

Around the equinoxes, such as 1965 and 2007, Uranus has a daily cycle more like an ordinary planet, and weather is more active.

Geocentric looping

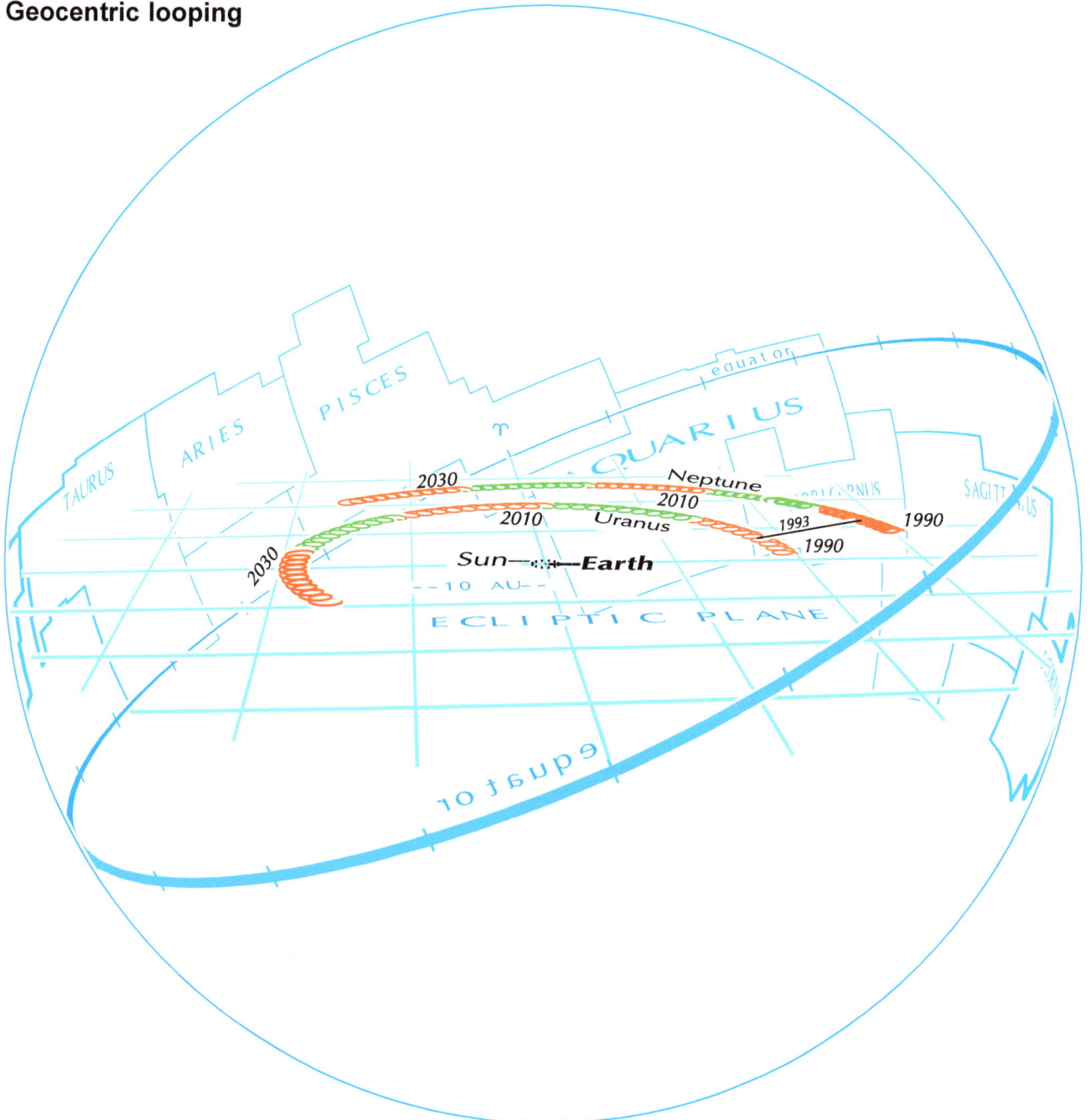

The dot at the center is the Earth, which in a geocentric frame of reference is held still, so that everything else appears to revolve around it. The Sun is shown as a little circle (exaggerated 10 times in size) at the beginning of each month, as it appears to travel around the Earth. This pre-Copernican conception of the cosmos still comes naturally to us.

The imaginary celestial sphere is 40 astronomical units in radius. The zodiacal constellations are drawn on the back, only, of the sphere. The viewpoint is from 15° north of the ecliptic plane, at longiftude 175° and at a distance of 120 a.u.

The paths of Uranus and Neptune are plotted for a half century, starting with 1990 so as to include the overtaking of Neptune by Uranus in 1993; a line connects them at their moment of heliocentric conjunction. The paths are drawn in red and green for alternate decades.

As seen from Earth, the planets appear to proceed in a spring-like path: each year they sweep inward and backward and then on forward. This explains the retrograde loops we see them trace against the starry background. The middle of each retrograde segment, when the planet comes nearest to us because the Sun is behind us, is its moment of opposition.

Pluto

Eccentric and inclined

Pluto's orbit is 248 years long—compare Neptune's 165. It is much more eccentric (out-of-circular) and inclined (tilted in relation to the ecliptic plane) than the orbits of any of the major planets.

Its average distance from the Sun is 39.4 astronomical units, which is well outside Neptune's 30. But at aphelion it is as far out as 49.3, and at perihelion it comes in almost half way, to 29.7. Thus it crosses Neptune's orbit, and is for 98 years nearer in. Yet it never comes close to Neptune or to any point in its orbit, because both Pluto's inward and its outward crossings are far north of the major planets' plane.

Pluto's visibility

Pluto is several thousand times dimmer than the dimmest stars you can see with the naked eye. Its magnitude ranges from 13.6 at perihelion to 15.9 at aphelion. The brightness varies by only about a tenth of a magnitude between opposition and other times of the year. Moving outward, it faded past 14 around 2010, and should fall past 15 around 2040.

Compare Neptune's magnitude of about 8. If we take the naked-eye limit to be about 5 (it varies from about 3.5 in city skies to 6.5 in excellent conditions), Pluto is from 3,000 to 23,000 times dimmer.

Pluto has been, since around 2000, in the southerly region of the sky through which the Sun moves in December. So the likeliest time for observation centers on July. The date of opposition falls in July from 2013 to 2029, after which it will move into August.

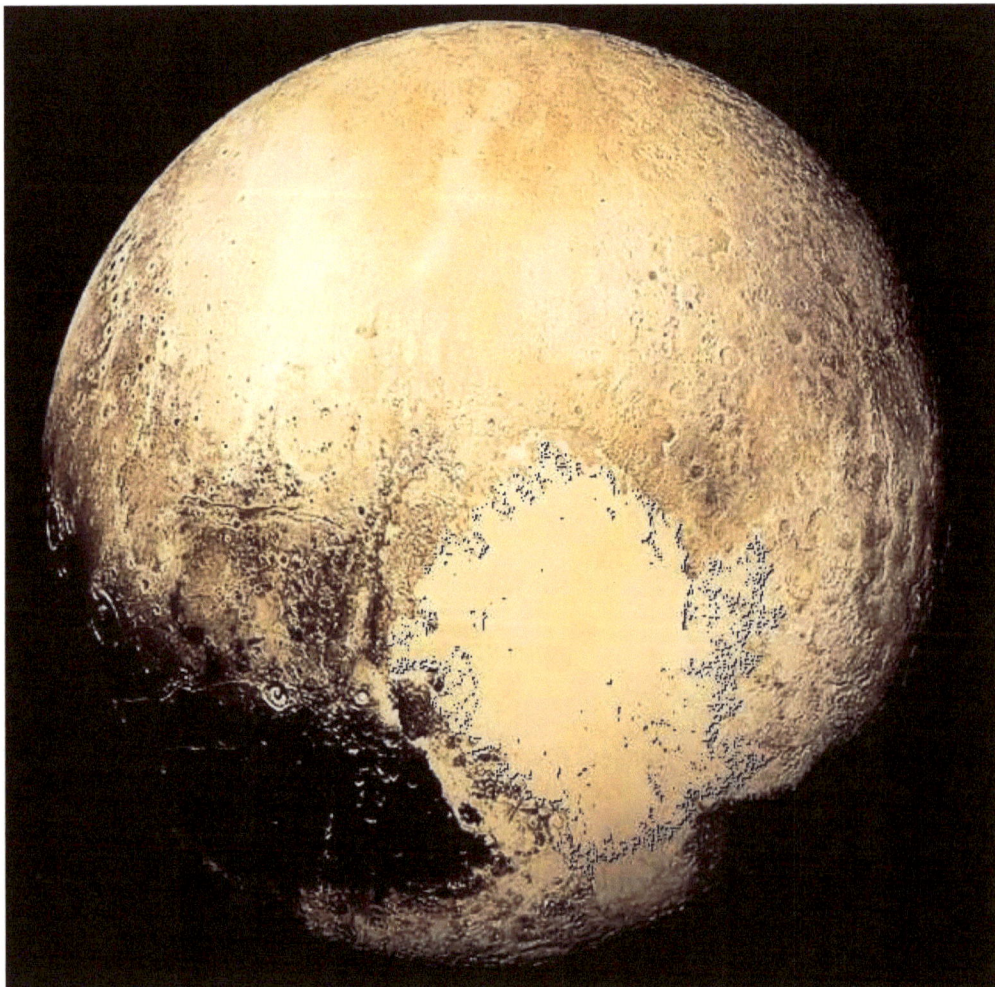

Pluto's course

Like Uranus, Pluto was discovered in Gemini—a northerly part of the ecliptic band, convenient for searches on the longer nights of the year, though rather full of Milky Way stars.

It passed less than 1 minute of arc south of 3.5-magnitude δ Geminorum (Wasat) on 1930 March 8. That star is almost on the ecliptic, and only half a year later (Sep. 9), Pluto was at the ascending node of its orbit through the ecliptic plane.

It was the outermost solar-system body then known, but was moving inward, toward its mean distance of 39.5 a.u. in 1937. In 1979 Feb. it became nearer in than Neptune, crossing very high over Neptune's orbit, because at around the same time, 1980 Feb., Pluto was at its highest latitude, about 17.2° or nearly 12 a.u. north of the ecliptic plane.

It moved through perihelion, its nearest-in point, in 1989. Then after 20 years as the "eighth planet from the Sun," it re-crossed Neptune's orbit (still well north of it) in 1999 Feb.

Because of Earth's own tiny circlings around the Sun, there was for a while still a time of each year—centered between our nearest time to Pluto's (its opposition) and our farthest time from Neptune (its Sun-conjunction)—when Earth was nearer to Pluto than to Neptune. The last such year was 2005.

For most of the time we have known Pluto, it has been north of the ecliptic, making an arch half way around the celestial sphere from ascending to descending node, with a 17°-high peak. It combed the mane, back, and tail of Leo; skimmed the northern fringes of other zodiacal constellations and southern corners of Coma, Boötes, and Serpens. Since 1988 it has been south of the celestial equator. In 2006 it entered the northwest corner of Sagittarius, across which it slants till 2023.

Throughout the northward arch of Pluto's course, the retrograde part of each year's path, centered on opposition, appeared (on the map of the sky) as a northward loop, because we were looking "up" to it from our ecliptic plane. The loop was fat when well north of the ecliptic, became slender until in 2013 it became a narrow spike. After that, the path became a flattened-S-like line which does not cross itself. When Pluto moves deep south of the ecliptic, the loop will re-open, southward.

From about 2000 to 2015, Pluto was crossing the Milky Way, a part of it not far from the galaxy's center, thronged—more thickly than the northern Milky Way in which it was discovered—with stars brighter than itself. In 2015-16 it had close encounters with the three stars π, o, and ξ of the "Teaspoon" of Sagittarius.

As it neared the ecliptic, Pluto became able to have close conjunctions with the Moon and planets. Between 2016 and 2024, it has 3 conjunctions closer than 1° with Mercury, 8 with Venus, 2 with Mars (0.013° on 2020 March 23), and 1 with Saturn.

Pluto's future travels

Pluto descends through the ecliptic plane on 2018 Oct. 25, half way around the sky from where it was discovered. It will crawl on out, through southern Aquarius, along the body of Cetus the Whale, to aphelion in 2114, 49 a.u. from the Sun; at its dimmest, it will be passing about 3° south of the red giant star Mira (whose magnitude can vary from below naked-eye to as high as 2). Pluto will come curving back through southern Taurus and northern Orion, to its region of discovery in 2178.

What will be the exact date when Pluto thus completes one orbit? Jean Meeus answered this question in an online discussion group in 2010, making use of Aldo Vitagliano's Solex software: on 2178 Jan. 22 Pluto will return to the same heliocentric longitude as in Tombaugh's photograph of 1930 Jan. 21.

So this one 248-year "year" of Pluto is uncannily close to an integral number of Earth-years!

The Neptune-Pluto standoff

Where did Pluto come from? How did it get to be in its peculiar orbit, slung in and out over Neptune's?

Many asteroids, and even more comets, have very eccentric and inclined orbits. There have been suggestions (Raymond Lyttleton in 1936) that Pluto is an escaped satellite of Neptune, perhaps knocked out of orbit by Triton, Neptune's currently largest satellite; or (Tom Van Flandern in 1979) that Pluto was a large comet, or a fragment torn from Neptune by a collision. Or that, like the similar Triton, it was an asteroid captured by Neptune. Pluto's orbit does not now come near Neptune's, passing north of it by a huge distance (8 astronomical units or Sun-Earth distances), but could it have been nearer in the past?

No: we have learned that this is impossible: Pluto has never, for millions of years, been near to Neptune.

The two are in a 3:2 orbital resonance. Their periods are 164.8 and 248 years. In about 495 years, Neptune goes around the Sun 3 times and Pluto 2.

When Pluto is at a perihelion, as in 1989, Neptune is roughly a quarter of the circle back from it. The next time Pluto is at perihelion, 248 years later, Neptune has gone around one and a half times and is far ahead of it.

Half way between two of these perihelia, Pluto is at aphelion, farthest out from the Sun; and, paradoxically, it is around this time that Neptune comes as near as it can. At the next Pluto aphelion, one of its orbits later and one and a half for Neptune, Neptune is on the opposite side of the Sun (near Pluto's perihelion), and they are as far apart as they can be.

Thus the nearest these two can come is over 17 a.u.—almost as great as the difference between Pluto's aphelion distance and Neptune's average distance (49 and 30 a.u.). Paradoxically, Pluto can come much nearer—11 a.u.—to Uranus than it can to Neptune!

The 3:2 orbital resonance—which we could perhaps even more clearly express as 1.5:1—is what keeps Pluto at great distance from Neptune; not the large inclination that causes Pluto's orbit to cross far north of Neptune's. The orbit could be flat, in the same plane as Neptune's, and the two bodies would still be kept far apart.

You may have noticed that the 3:2 relation between the periods is not, at present, exact. We should have called it a relation between the *mean* periods. All the orbital elements of all planets change slightly over time, being subject to continual gravitational "perturbation" by each other. However, in the 3:2 resonance the changes are self-correcting.

Pluto's orbital period is at present slightly longer than 1.5 times Neptune's. So each time Neptune overtakes Pluto it does so slightly sooner; it is pulled slightly forward, and Pluto slightly backward. Neptune's orbit is slightly shortened and Pluto's lengthened. At each repetitions, the effect becomes slightly smaller, until it reverses; they drift back through the exact 1.5 ratio and out on the other side. This cycle takes about 20,000 years. And it will have been going on for millions. The 3:2 state of the orbits is very stable.

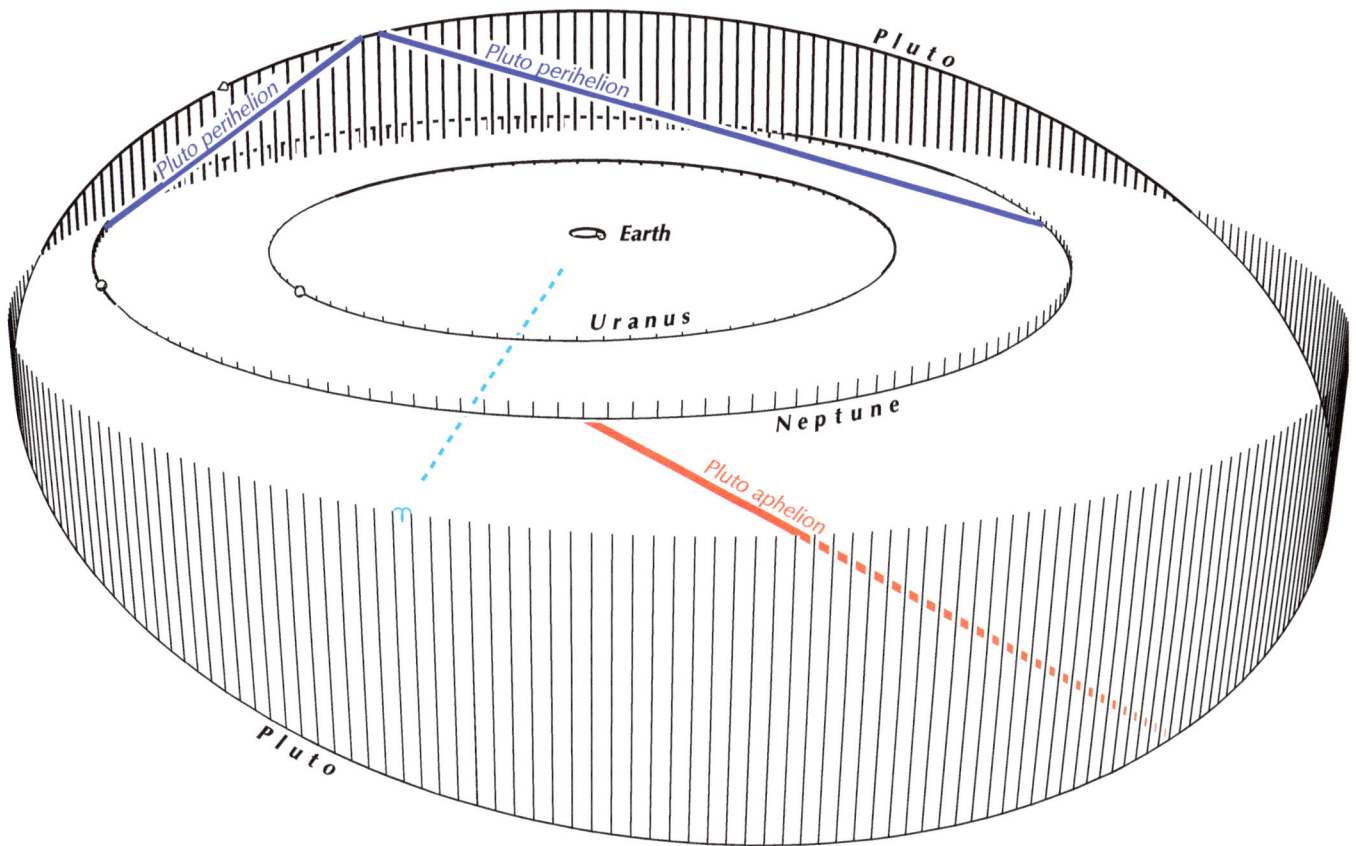

Orbits of Uranus, Neptune, and Pluto, with stalks to the ecliptic plane at yearly intervals. The blue lines connect Pluto to Neptune at the times of two of Pluto's perihelia. The red line connects them at one of Pluto's aphelia; a few years after this, Neptune will be as near to Pluto as it can get. At the Pluto aphelion an orbit later, Neptune will be an orbit and a half away, on the opposite side of the Sun. The dashed line shows the vernal equinox direction.

Plutinos

From 1993 on, many smaller bodies have been found that, like Pluto, are in 3:2 orbital resonance with Neptune. They are called Plutinos. Some have received names which, like Pluto's, resound with the underworld—Orcus (old Latin counterpart of Hades), Ixion (the evildoer eternally bound to a fiery wheel).

The Plutinos form the inner or Neptune-crossing part of the Kuiper Belt. Their orbits have various inclinations and eccentricities, one of them coming in as far as half way between Neptune and Uranus. And they are subject to the same oscillation around the exact resonance with Neptune—and to very slight gravitational influence by the largest of them, Pluto. If they all, including Pluto, had zero inclination and zero eccentricity and stayed in the exact resonance, all would be at a point, about 9.3 a.u.

out beyond Neptune. They are not, of course, nor do they form a cloud around Neptune, but are a scattered "dynamic group," linked by the nature of their orbits.

Besides the 3:2 resonance, there are others, such as 2:1, in transneptunian zones farther out, acting less powerfully and with fewer bodies obeying them.

How did these resonances come to be? This gets into mathematical studies of the formation of the planets from the primordial disk of matter around the Sun. Computer models have suggested that the major planets changed their positions. Neptune migrated outward; when it approached the inner edge of the Kuiper Belt of small bodies, its gravity captured at least one (Triton), threw others into chaotic orbits, and forced others into resonance with itself.

Pluto's much-more-than-sidewise rotation

Earth and Mars have days of roughly the same length, Mercury and Venus take an extraordinarily long time to rotate, the four giant planets rotate surprisingly fast. Pluto is slow. Though it is far smaller than Earth, it takes about 6.4 days to turn around—its day is nearer to a week of ours.

Because of its slow rotation, Pluto has essentially zero "flattening": it is not compressed from pole to pole, as are Earth, Mars, and the giant planets, especially low-density Saturn.

And Pluto rotates "on its side," like Uranus. We state the angle of inclination of a body's axis in relation to the plane of its own orbit (not to the Earth's ecliptic or equator). In these terms, Pluto's odd inclination is more extreme than Uranus's: nearly 123° (as against Uranus's nearly 98°).

When we say planets rotate "on their sides," this is a status only roughly intermediate between prograde and retrograde. You cannot be extremely middle; the most middle you can be is exactly so, which a planet could achieve only by having an axial tilt of 90°—flat along its orbit. Uranus errs from that by 8°, Pluto by 33°. The axial tilts of planets, including Earth, vary over very long times; Pluto's, it is calculated, varies between 102° and 126°.

So we have the same dilemma: which of its poles is "north"?—equivalent to deciding whether its rotation is prograde (counterclockwise as seen from the north, like most of the planets) or retrograde (the untypical way).

One of the poles of Pluto's axis aims at a point in the sky at right ascension 313.02° (20h 52m), declination -9.09°. (So said the 4th edition, 2000, of *Allen's Astrophysical Quantities*.) This point is in western Aquarius, a little northeast of the naked-eye star Epsilon Aquarii, also called Albali (Arabic, "the swallower"),

This point, like the corresponding one for Uranus, is south of the celestial equator but north of the ecliptic, which is why it was regarded as Pluto-north. It is about half way between those planes, whereas the Uranus pole is less far north of the ecliptic. As for Uranus, the official stance was that the rotation is retrograde. The opposite pole was at 8h 52m, +9° (at the southern edge of Cancer, just north of the "head" of Hydra).

Not surprisingly, it was difficult to fix exactly the point on the distant little body around which it slowly rotates. Measurements in the New Horizons era have found this pole to be about 15° farther north. That puts it in Delphinus (close to the border with another little constellation, Equuleus)—north not only of the ecliptic but of the celestial equator. So it is more inarguably Pluto's north rotational pole, right?

No. Evidently it was realized that in describing Pluto's details it makes more sense to picture it rotating directly, as Earth does; in other words, to use the convention Meeus recommended. So the opposite pole, from which Pluto would be seen rotating counterclockwise, becomes the northern. Maps of Pluto show the corresponding pole at the top. It points at right ascension 133° (8h 52m), declination -6.2°, in western Hydra (south of the "head" of the water-serpent and west of Alpha Hydrae, Alphard). It is a prime example of a "north" pole that is south not only of our equator but of our ecliptic, for good reason.

Pluto passed one of its equinoxes (its equator in line with the Sun) in 1989. This almost coincided with Pluto's perihelion, also in 1989. In other words, the inner and faster part of Pluto's orbit, centered on perihelion, coincides with the span between the two solstices (about 1914 and 2030) or times when the poles face the Sun. This seems hardly likely to be a coincidence.

From 1989 on, Pluto's northern hemisphere has been tilting increasingly toward us. So for New Horizons, as it whizzed by in 2015, the southern hemisphere was mostly hidden in darkness.

Pluto's orbit slopes across the ecliptic at 17°, but that is not the reason for the tilt in relation to its own orbital plane. As it continues outward and southward along this plane, through Sagittarius and southern Capricornus, Pluto continues to keep its south pole aimed 30° north of its plane (at Delphinus, as seen from Pluto though not from our viewpoint).

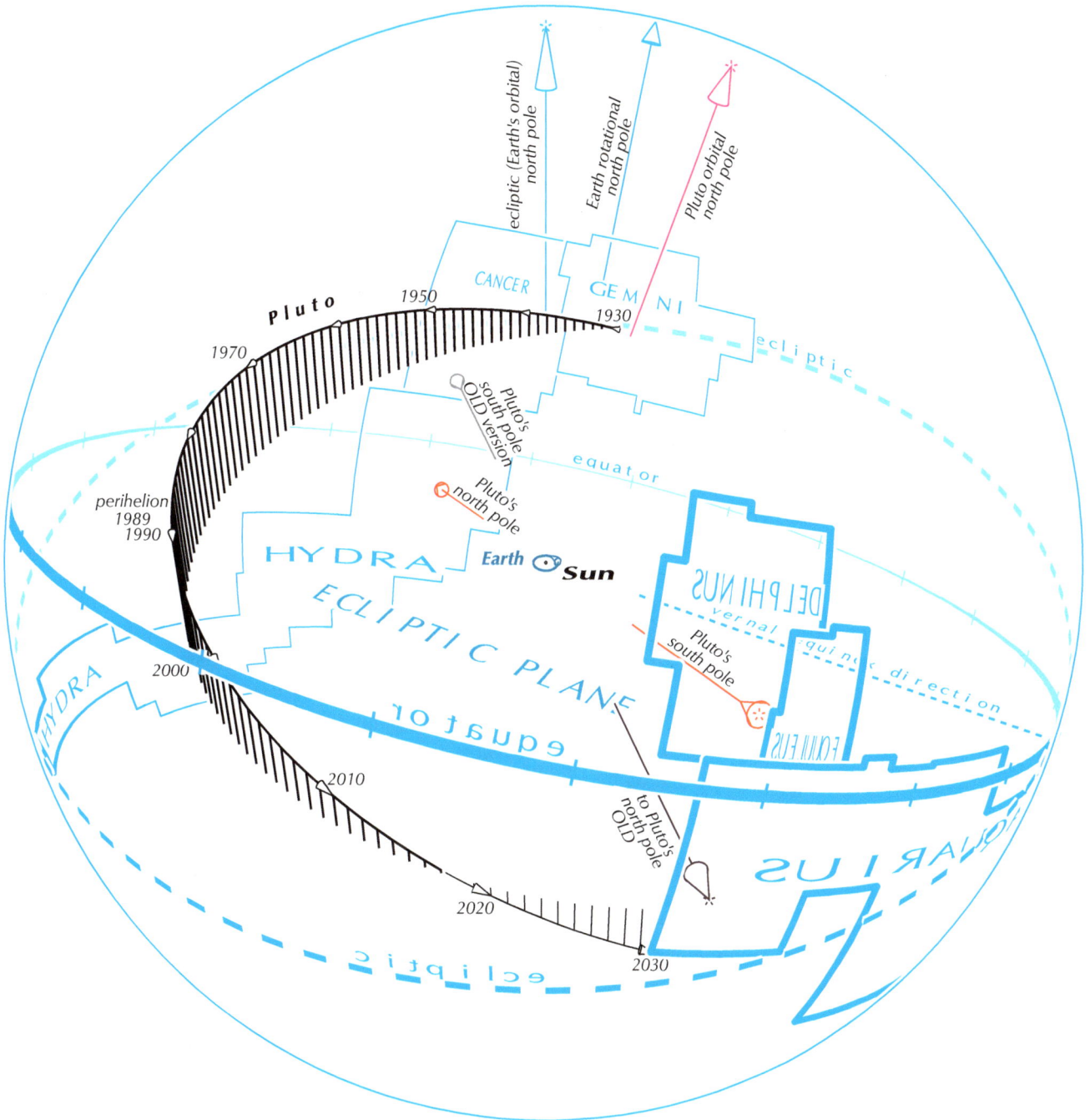

Pluto's trajectory from its 1930 discovery, in Gemini, to 2030. The radius of the imaginary celestial sphere is 42 astronomical units (to fit in Pluto's 1930 and 2030 distances of about 41 and 37 a.u.). Arrows show the directions (from the Sun) to the two poles of Pluto's rotational axis, in two versions. In the earlier (gray), the north pole was in Aquarius and the rotation was considered retrograde; in the current version (red), the north pole is in Hydra and the rotation prograde. Imagine the tiny globe of Pluto at each point along its path, with a miniature red arrow through it pointing parallel to the large one.

Also shown is the north pole of Pluto's orbital plane, the north pole of the ecliptic system (that is, of Earth's orbit and, roughly, of the solar system in general), and Earth's equatorial (rotational) north pole.

A little sparkle surrounds the tip of an arrow if it "pricks" the nearer side of the celestial sphere. The view is from ecliptic latitude 35°, longitude 300°, at a distance of 120 astronomical units from the Sun, whose size is exaggerated by 20.

Pluto's family

Pluto has five known satellites. It has no rings as the giant planets do.

The first satellite, Charon, now known to be large, was revealed in 1978 as a tiny excrescence on the indistinct and almost as tiny image of Pluto, in photographs made in earlier years at the Lowell observatory in Flagstaff. The photos had been labeled "poor" because the bulge was assumed to be a defect, but James Christy, at the U.S. Naval Observatory in Washington, noticed that stars in the same plates did not show the distortion and that it moved around Pluto's edge. The guess that it was something real, a satellite, was confirmed by calculating an orbit for it, from which could be predicted a series of eclipses of satellite and Pluto by each other between 1985 and 1990; and these were observed.

The other four satellites were discovered by members of the team preparing the 2015 New Horizons mission to Pluto: Nix and Hydra in 2005, Kerberos in 2011 and Styx in 2012. They are from 2 to 4 times farther out than Charon, and only from 19 to 50 kilometers, so that unlike Charon they have not contracted into spherical shape.

And the small satellites, unlike the synchronous Pluto-Charon pair, rotate quickly.

Grim ferryman

The satellites of Pluto, god of the dead, have received suitably sepulchral names.

In Greek mythology, Charon was the ferryman who carried the souls of the dead across the river Styx into Hades. Kerberos, or Cerberus in Latin form, was the fearsome three-headed dog who guarded the gate of the underworld. Hydra was the many-headed water serpent whom Hercules had to slay in one of his twelve Labors. *Nyx* is Greek for "night," but, since that name was already applied to an asteroid, was altered to *Nix*, which happens to be Latin for "snow," and slang English for "negative."

Satellite Charon is so closely associated with Pluto that it might appropriately have been named for Persephone, the consort of Hades (Pluto), who abducted her from her sunny life in the overworld. That was suggested, but James Christy, the discoverer, had already picked Charon because it could allude to his wife, Charlene, nicknamed Char, with initial consonant sounded *sh*, as if the word were French. Hence, astronomy insiders pronounce it that way, to show that they know. I'm not sure I would want to identify my wife with the janitor of hell.

Sources such as dictionaries may give the name's first vowel as that in *share*, and the second as the reduced vowel (schwa) as in *reckon*. In my *Albedo to Zodiac*, I explain some of the linguistic history behind our choices of pronunciation, but I favor a tolerant approach, and I say the two syllables of *Charon* like *car* and *on*.

The Pluto-Charon embrace

Dwarf planet Pluto can scarcely be described without including its close partner Charon.

Earth and its Moon are sometimes said to be, or almost to be, a double planet, because the Moon is larger in relation to its planet than other satellites are. Except for Charon, which in this respect much exceeds the Moon, so that Pluto and Charon have a stronger claim to be a binary system—a double dwarf planet.

Charon is only 12th in size among the solar system's natural satellites, but in relation to its primary it is far larger than any other. It is just over half as wide as Pluto (diameters 1,212 and 2,376 kilometers). Its mass is about 1/8 of Pluto's. No satellite of a major planet comes close to these proportions. The Moon has about 1/81 of Earth's mass, Ganymede 1/12300 of Jupiter's, Titan 1/4150 of Saturn's, and Triton 1/4800 of Neptune's.

And Charon is near to Pluto. In kilometers, their radii are about 1,200 and 600, and the distance between their centers is only about 19,600,

In all major-planet-plus-satellite systems, the barycenter or center of mass, around which the members of the system revolve, is inside the planet. For Earth and Moon, for example, the barycenter lies 1,710 km under Earth's surface. But for the Pluto-Charon system, the barycenter is 2,110 km from Pluto's center—out in space beyond its 1,200 radius.

Many satellites near to their planets, including the Moon, are tidally locked, that is, keep the same face toward their planet. In this case, there is mutual locking: Pluto, too, keeps a same face toward Charon. So it is not just Pluto that rotates in about six days, but Charon too.

In the mapping system used for the surface of Pluto, the prime meridian (like the Greenwich meridian for Earth) is the longitude facing Charon

Charon, unlike Pluto, has no atmosphere. Its materials are similar to Pluto's, but its surface is darker and duller, made of water ice with some ammonia. It has mountains and deep canyons, formed more than four billion years ago before it cooled and ceased to be geologically active.

There are material differences on Pluto and Charon caused by each other. Some substances are more abundant on Pluto's away-from-Charon side. From Pluto's thin atmosphere have flowed gases that have helped to form Charon's reddish northern cap of compounds called tholins, which may be involved in the origin of life.

One hypothesis about the origin of Charon is that it was born from a huge impact on Pluto, and the smaller satellites formed of debris left over from the same collision. Another is that Charon originated as a transneptunian that collided with Pluto gently enough to go into orbit, without being shattered and re-formed like the Moon.

Pluto's painted deserts

At first a mere dot in Clyde Tombaugh's blink comparator, Pluto turns out to be a dwarf planet of great interest, full of detail. It has peaks, canyons, pits, wrinkles, sheets and polygons of ice; it exceeds most planets in contrasts of brightness and of color, which ranges from black to reddish brown, orange, yellow, and white.

Its diameter, currently fixed at 2,376 kilometers, is far smaller than was expected at its discovery, though larger than had been estimated in recent decades. Its mass, too, when the discovery of Charon enabled it to be determined, is far less than had been thought. In both size and mass, it is smaller than seven satellites: Ganymede, Titan, Callisto, Io, the Moon, Europa, and Triton. Its mass is 18 percent of the Moon's, 0.22 percent of Earth's. Its surface area matches that of Russia.

New Horizons sent back a wealth of surprising information, which will take years more to process. It found evidence of geological activity, which had been assumed to cease long ago as small Pluto cooled.

From a distance the outstanding feature was a light-colored heart-like shape, nearly a thousand miles wide, given the name Tombaugh Regio. (Names may change, being not official until approved by a committee of the International Astronomical Union.) Closer up, the western and brighter half of the "heart" was found to be a bowl, Sputnik Planitia, probably made by an enormous impact less than 100 million years ago. It is without impact craters, which are an indication of age, and is being continuously renewed: it is divided into polygonal cells with trenches around them, as by convection from underlying warmth. Bordering it on the southwest is the largest dark region, at first called "the Whale," then Cthulhu Macula

(from a being in an H.P. Lovecraft story; *macula* is Latin, "spot of dirt"); it is heavily cratered, and thought to be draped with a "tar" of hydrocarbons called tholins. On west along Pluto's equator are the Brass Knuckles, a row of large dark holes.

Pluto's surface is a patchwork of ices, chiefly of nitrogen, also of methane, carbon monoxide, and water. Only the water ice is stiff enough to poke up through the sloppy nitrogen, forming tall mountains. There are two broad water-ice volcanoes that may have been active not long ago. There are long systems of faults, regions of parallel or forking "slices," mountains capped with methane snow, glaciers sliding from mountains onto plains, and thousands of pits from which nitrogen has sublimated.

Beneath the mainly nitrogen-ice skin is the mantle, of water ice, forming most of Pluto by volume. At the warmer bottom of this may be a liquid-water ocean. Then the core, of silicate rock, making more than half of Pluto's mass. The core does not act as a dynamo: unlike the major planets, Pluto has no magnetic field.

But, unlike Mercury or the Moon, it has an atmosphere, mostly of nitrogen, like Earth's, also methane and carbon monoxide. It is thin, but reaches out hundreds of miles; has no clouds, but many haze layers, which give a blue sky.

The atmosphere is escaping, though more slowly than had been expected. And some of the surface is being blown off by the solar wind. (If Pluto were nearer in, would it develop a gas-and-dust tail like a comet?) With increasing distance from the Sun, some of the atmosphere should freeze onto the surface, but this seems to be happening slowly; some atmosphere may survive throughout the orbit.

Seasons in the world of Hades

We can describe Pluto's "year" by comparing it with Earth's. Earth's rotational day is 24 hours long; Pluto's is 6.4 Earth-days long. Earth's around-the-Sun year is 365 days long; Pluto's is 248 Earth-years long. Earth's distance from the Sun varies little; Pluto's distance at aphelion is almost twice its distance at perihelion. Earth's rotation axis stands not at 90° but at about 66.6° to its orbital plane around the Sun; Pluto's poles point only 33° away from its orbital plane.

Earth's perihelion is rather near to one of its solstices, but this has little effect. Remarkably, Pluto's equinoxes almost coincide (on a Pluto time-scale) with perihelion and aphelion. The north-hemisphere spring equinox was in 1989, when the planet was at perihelion, most Sun-warmed and traveling fastest.

Suppose we broaden "summer" to mean the whole more-sunward-facing half of the "year" for one of the hemispheres, between the equinoxes and centered on the solstice; "winter" likewise to mean the other half. For Pluto, each of these seasons is 124 Earth-years long. Since they stretch from equinox to equinox, each hemisphere receives the same total of sunlight, though for the northern hemisphere summer begins more rapidly and its warmest part is earlier; vice versa for the other hemisphere. These prolonged seasons are extreme: the half enjoying summer gets relatively much warmer and the half suffering winter gets much colder.

From 1989—perihelion and equinox—it is summer for the northern hemisphere (the hemisphere that New Horizons could mainly see).

Think what it would be like if not an equinox but a solstice were to coincide with perihelion—Pluto being turned a right-angle. At perihelion in 1989, not the equator but one pole would point as near as it could toward the Sun, in the middle of an optimal summer. 124 years later, that hemisphere would have the coldest possible winter, facing away when most distant. Instead of symmetry between the seasons for the two hemispheres, there would be maximal imbalance.

Around each of Earth's poles there is an Arctic zone, 23.4° in radius, that gets at least some 24-hour sunlight in the half of the year centered on the summer solstice, and at least some 24-hour darkness in the half of the year centered on the winter solstice. The much wider zone, from 66.6° north to 66.6° south (about three quarters of Earth by latitude, far more by area), gets at least some sunlight, as well as some night, for at least a part of every day in the year.

But on more-tilted Pluto the polar zones are larger, 57° in radius, and these are the zones that get times of uninterrupted sunlight in summer and darkness in winter—many-year-long uninterrupted times for parts nearer to Pluto's poles. Each pole actually gets more total time in sunlight than the equator.

Deeply changing seasonal conditions have to affect the chemicals on Pluto—their freezing onto or blowing off the surface, their piling or crevassing, their migration from place to place by storm or glacier.

Charts

These charts are ecliptic-based, to save vertical space. The more familiar lines of the equatorial system (right ascension and declination) are shown, sloping and curving in relation to the ecliptic system.

Long-term chart for Uranus and Neptune   From 2010 to 2030.

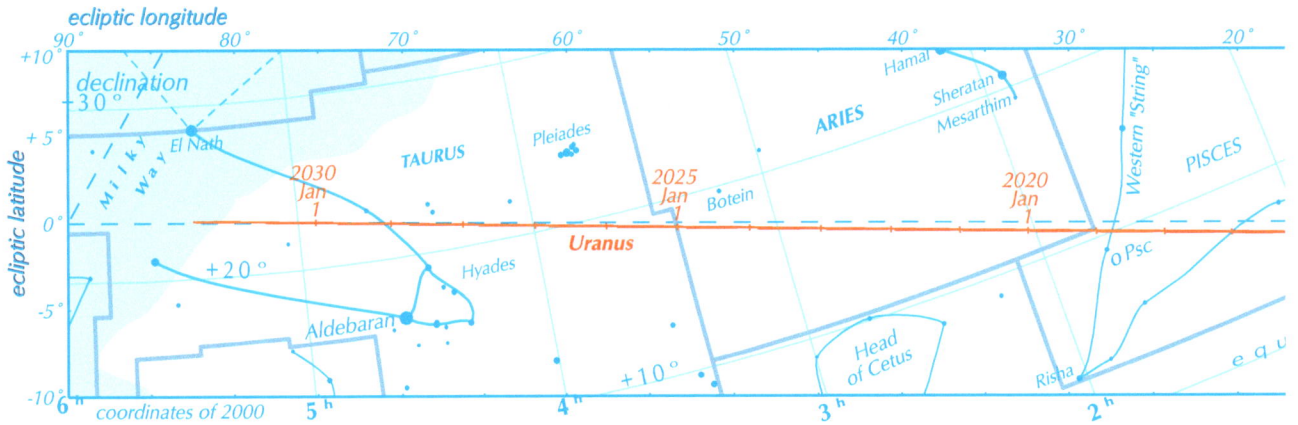

Long-term chart for Pluto

From 1930 (discovery) to 2030. To make the separation of years clearer, the tracks for each 5th year are in darker color. The Sun's course in November and December is shown as a

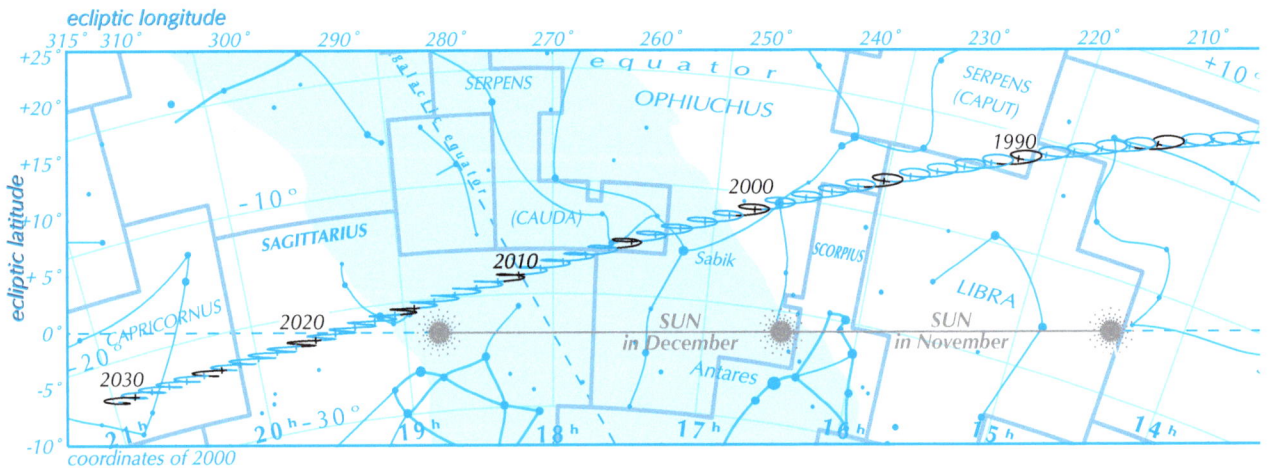

Charts for Uranus and Neptune

The positions of Uranus and Neptune at the beginning of each month are indicated by circles, proportional in size to their brightness, on the same scale as the filled circles for the stars. Thus you can find them by comparing with neighboring stars.

Labels point to January 1 of each year, but not the other months, to avoid crowding. Also labeled are the dates of opposition.

Because Uranus and Neptune have slow and low-inclination orbits, the paths in successive years, and especially the forward and backward parts, almost overlie each other and are difficult to distinguish. To alleviate this, a reddish color is used for the retrograde (backward, westward, left-to-right) parts of the paths. These are the times when the planets are nearer to us, so they are brighter, and the retrograde tracks appear slightly farther from the ecliptic. Opposition happens in the middle of each retrograde track.

At dates when Uranus or Neptune is passed by another planet, a green line connects them at

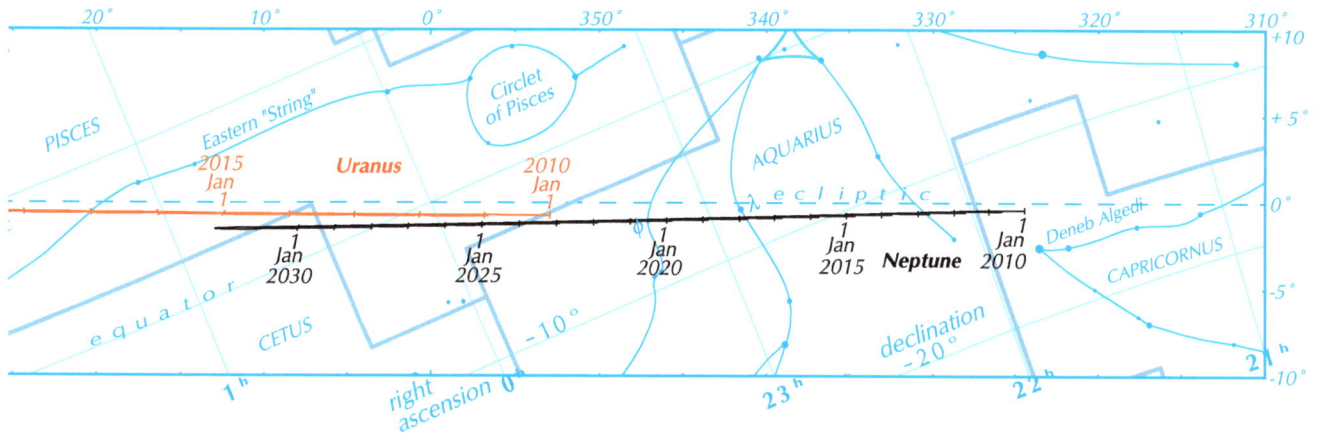

reminder that, in our times, Pluto's oppositions have been in July, moving toward August, so its Sun-conjunctions are in December, moving into January.

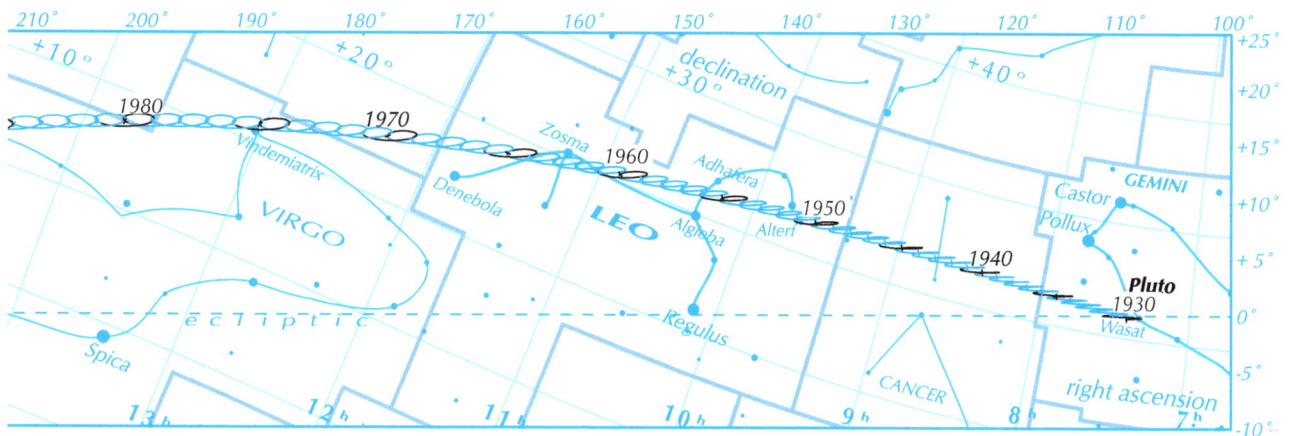

the instant when they appear closest to each other. This moment of closest approach may be loosely called a conjunction, but is more strictly an appulse. It happens at a time differing somewhat from the conjunction in longitude (when the connecting line would be vertical in our chart) and the conjunction in right ascension (when the connecting line would be parallel to the lines of right ascension). The other planet, being much nearer to us, is usually faster-moving on the map of the sky. Its course over the day of the appulse is drawn with an arrow.

Sometimes the passing planet is outside the top or bottom of the chart. Not shown are appulses wider than 5°, or at elongations of less than 15° (in bright sky near the Sun).

Stars have numbers, and the brighter ones Greek letters, in sequence in their constellations; for instance Aldebaran is also α (alpha) Tauri or 87 Tauri. Some deep-sky objects, such as nebulae and star clusters, are marked with their numbers in the NGC (*New General Catalogue*, 1888).

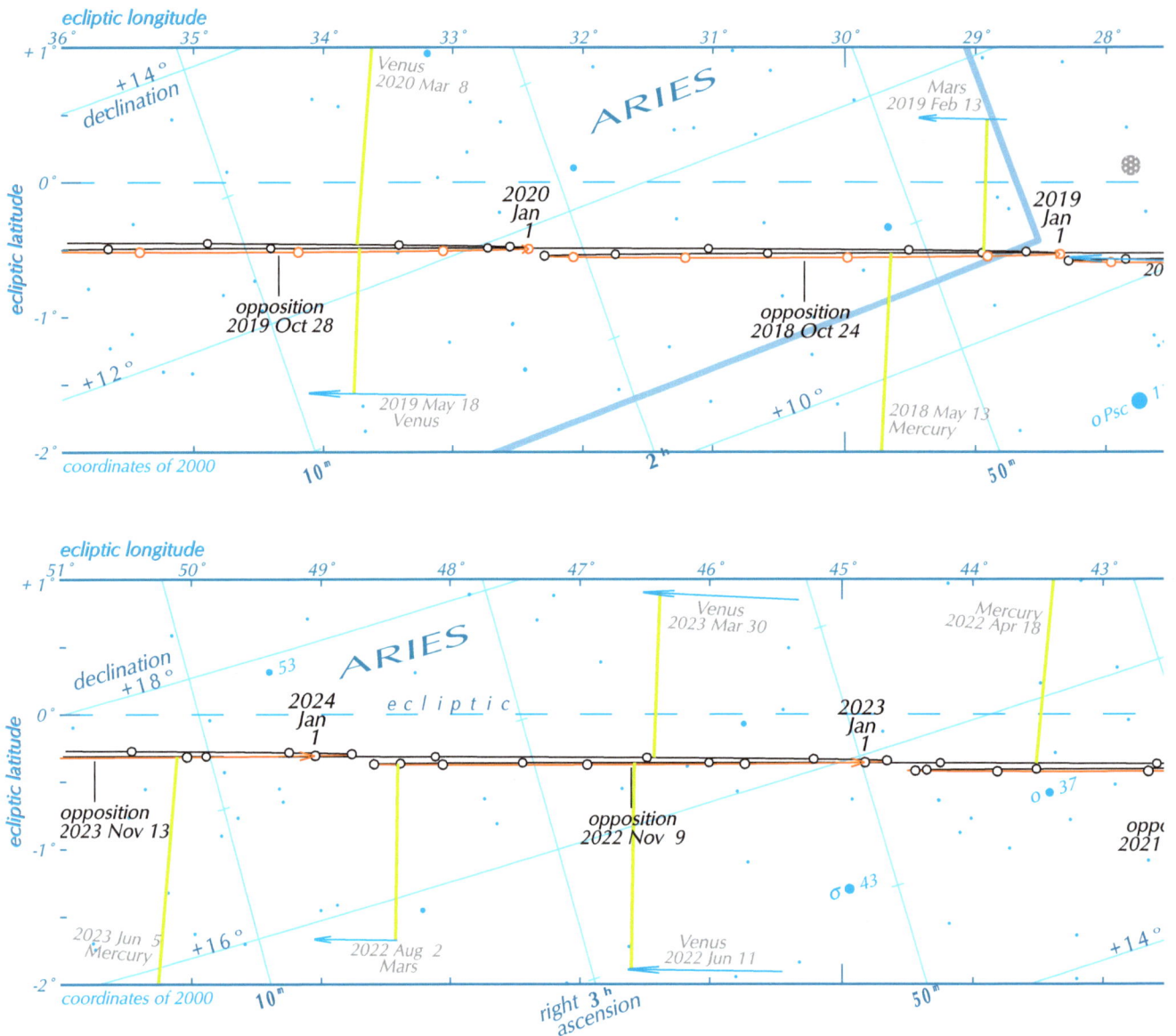

Uranus crossed the celestial equator into the sky's northern hemisphere in 2011-2012. It is in the sky region that, for Earth's northern hemisphere, begins to rise into view in August evenings and is at its modest highest over the southern horizon when the planet is at opposition in October.

This date of opposition moves into November in 2021, and into December in 2028. The opposition-centered above-the-horizon span of hours grows longer as Uranus climbs higher north.

Its path lies south of the ecliptic, but rising slowly toward it, because Uranus was at greatest latitude south in 2007. The latitude was 0.77° in 2007, 0.60 at the beginning of 2017, will be

0.44° at the end of 2020, and 0 on 2029 May 10—ascending node. So, during this arc of time, the retrograde loops, always narrow, become progressively even narrower, to be straight lines in 2029.

Aphelion was in 2009 and perihelion will be in 2050, so Uranus is becoming each year slightly nearer in, at a slowly increasing rate. Its distance from us at opposition decreases from about 18.9 a.u. in 2017 to 18.7 in 2021 and 18.2 in 2029.

The apparent width of its disk at opposition grows from 3.7″ in 2017 to 3.8″ in 2023 and 3.9″ in 2030, and its brightness from magnitude 5.7 in 2017 to 5.6 in 2021 and 5.5 in 2029.

Chart 1 (Pisces):

28° 27° 26° 25° 24° 23° 22° 21° 20° +1°

PISCES

Mercury
2017 Mar 26

ecliptic

Mars
2017 Feb 26

Uranus
2017
Jan
1

2019
Jan
1

2018
Jan
1

0°

2018 Mar 29
Venus

opposition
2017 Oct 19

o Psc 110

2017 Jun 3
Venus

+8° declination

−1°

right ascension

40ᵐ 30ᵐ 20ᵐ −2°

Chart 2 (Aries):

43° 42° 41° 40° 39° 38° 37° 36° 35° +1°

γ
8

ARIES

Mars
2021 Jan 20

+0ᵐ

ecliptic

Uranus
2021
Jan
1

2022
Jan
1

0°

o 37

opposition
2021 Nov 4

opposition
2020 Oct 31

−1°

+14°

+12°

40ᵐ 30ᵐ 20ᵐ −2°

magnitudes
4
5
6
7
8

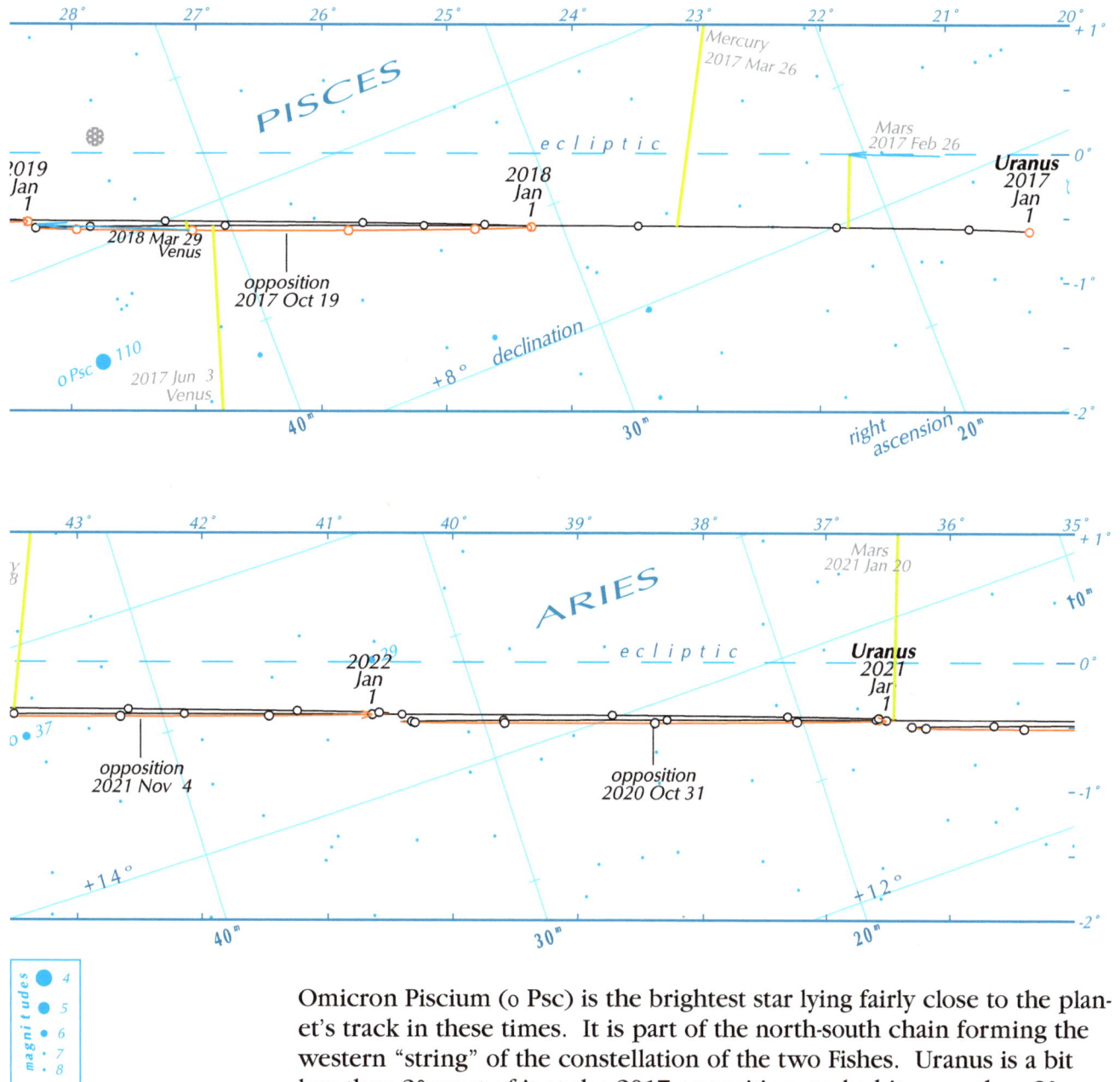

Omicron Piscium (o Psc) is the brightest star lying fairly close to the planet's track in these times. It is part of the north-south chain forming the western "string" of the constellation of the two Fishes. Uranus is a bit less than 2° west of it at the 2017 opposition, and a bit more than 2° northeast of it at the 2018 opposition. The star's magnitude of 4.3 is a step and a half above Uranus's 5.7, so comparing the two could be an interesting demonstration of how Uranus hovers near the naked-eye limit.

On 2018 Jan. 2, Uranus becomes stationary just short of reaching the Pisces-Aries border. Re-advancing, it reaches Aries in May, but crosses back in December, and only in 2019 Feb. arrives definitively into Aries.

At the end of August 2021, Uranus slides eastward just north of Omicron (o) Arietis; retreats even more closely past it in October; and at opposition in early November is about a degree west of it. The magnitude of this star, 5.8, is almost the same as the planet's 5.6 at opposition.

58

Top chart

ecliptic longitude

66° 65° 64° 63° 62° 61° 60° 59° 58°

+1° — +22° declination

TAURUS

Venus 2026 Apr 24

2027 Jan 1

202 Ja 1

0°

14

2026 Jul 4 Mars

opposition 2026 Nov 25

opposition 2025 Nov 21

50

-1° — +20°

2025 Jul 4 Venus

37
39

43
ω

coordinates of 2000 10ᵐ right 4ʰ ascension 50ᵐ

Bottom chart

ecliptic longitude

81° 80° 79° 78° 77° 76° 75° 74° 73°

+1° — declination +24°

13

99

NGC1746

TAURUS

2030 Dec 12 opposition

2029 Dec 8 opposition

0°

1 Jan 2030

18

-1° —

2030 Jul 9 Venus

+22°

15

ι 12

2029 Jun 24 Mercury

2028 Ju Ver

20ᵐ coordinates of 2000 10ᵐ right 5ʰ ascension 50ᵐ

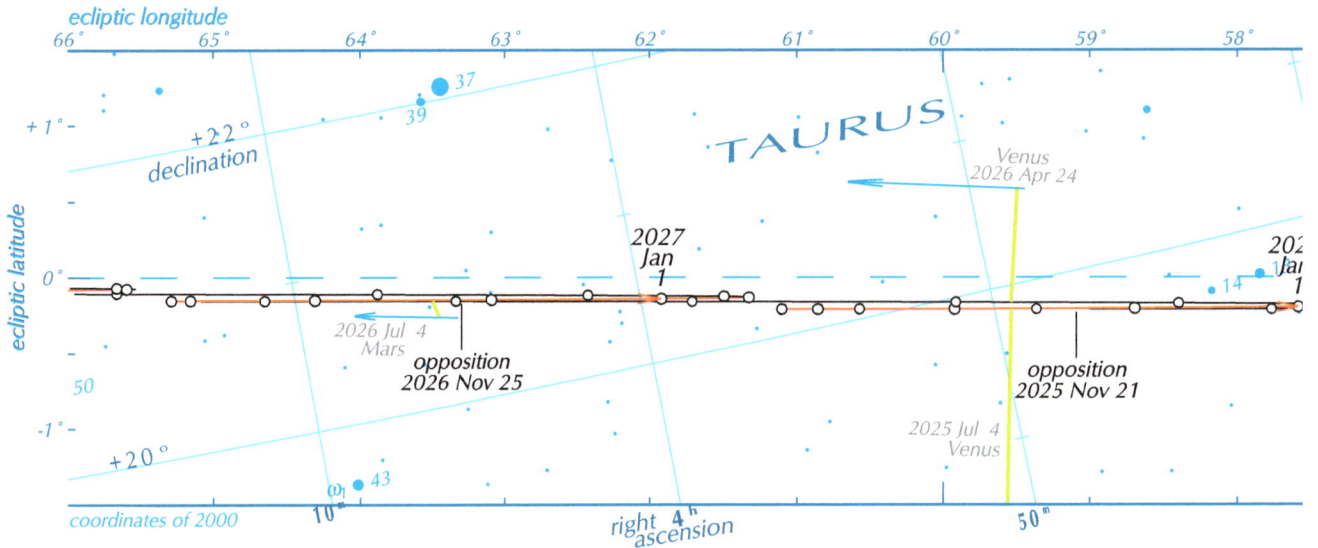

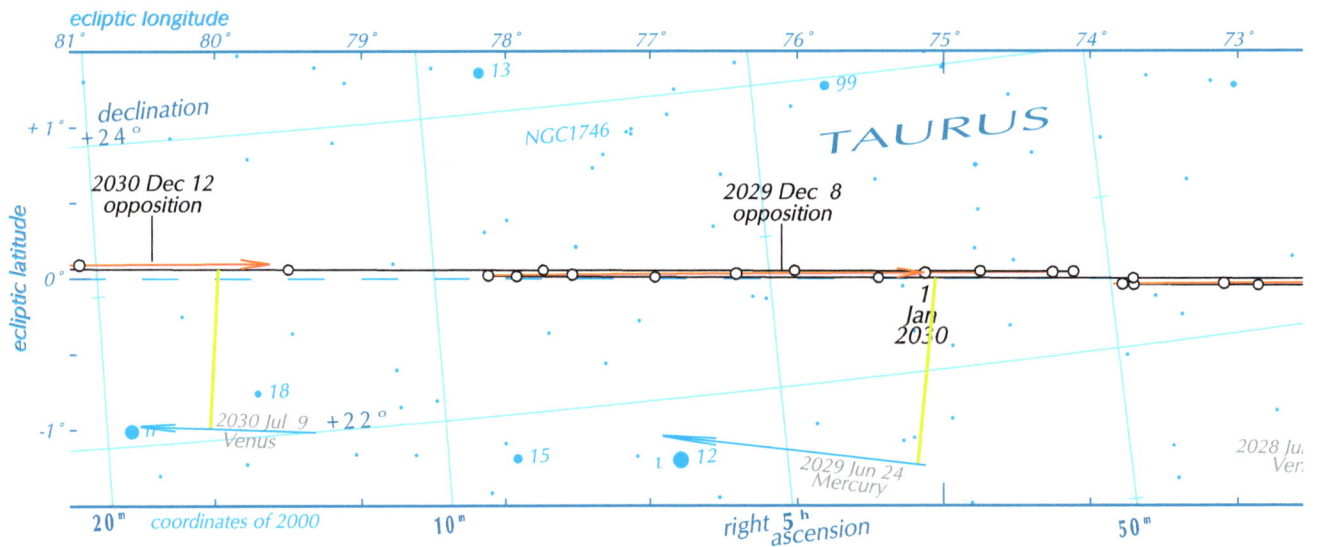

NGC 1746 in Taurus, about 5° north of Uranus's 2029 track, is in the *New General Catalogue* because it was described in 1863 as a star cluster by Heinrich d'Arrest—he who had been Galle's assistant at the first sighting of Neptune in 1846. It was shown in 1998 to be not a real cluster but a chance grouping of stars at different distances.

After many years of travel across star-poor constellations, Uranus from about 2030 begins to cross the somewhat more crowded background of the northern Milky Way in Taurus and Gemini.

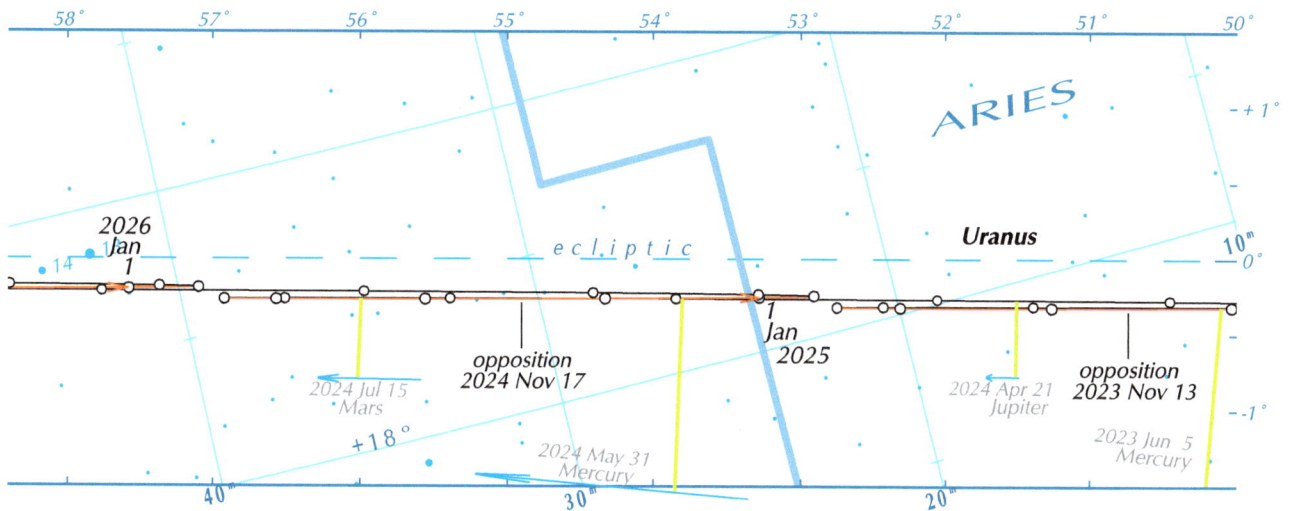

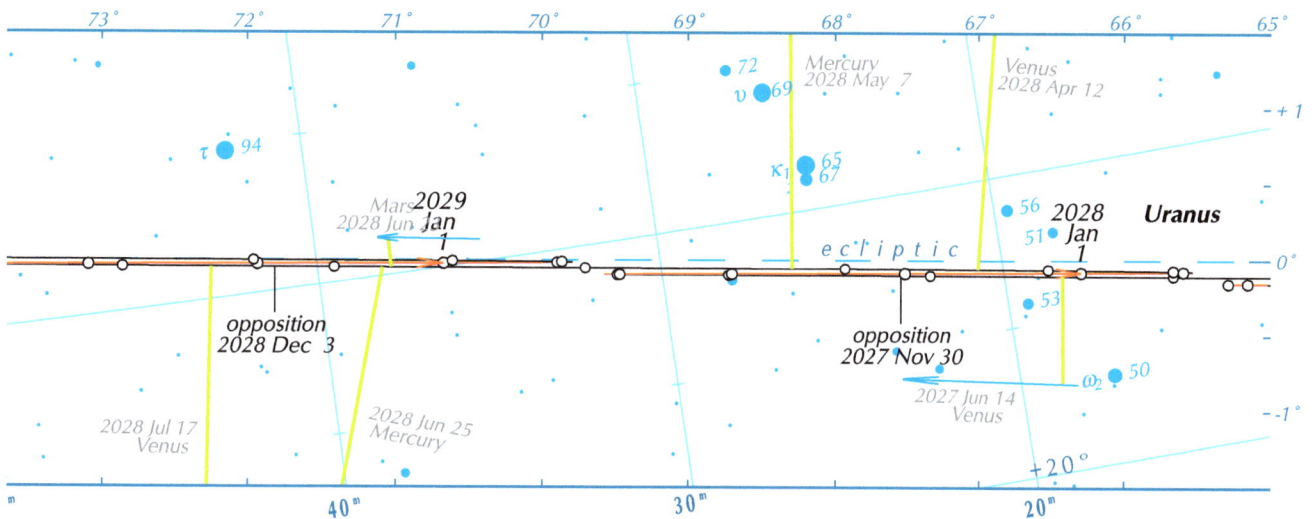

On 2023 Aug. 28, Uranus becomes stationary half a degree short of the end of Aries. After moving back through the opposition of Nov. 13, it starts forward again on 2024 Jan. 27; enters Taurus in June; after moving about 3 degrees in, and back through the next opposition (2024 Nov. 17), it dips back into Aries at the end of Dec. 29; ends its backward movement on 2025 Jan. 30; and commits to Taurus on 2025 Feb. 3.

In 2026 Uranus lies just over 4° south of the Pleiades. In 2027 and 2028 it is about as far north of the more scattered Hyades, and 5° north of

Charts—Neptune

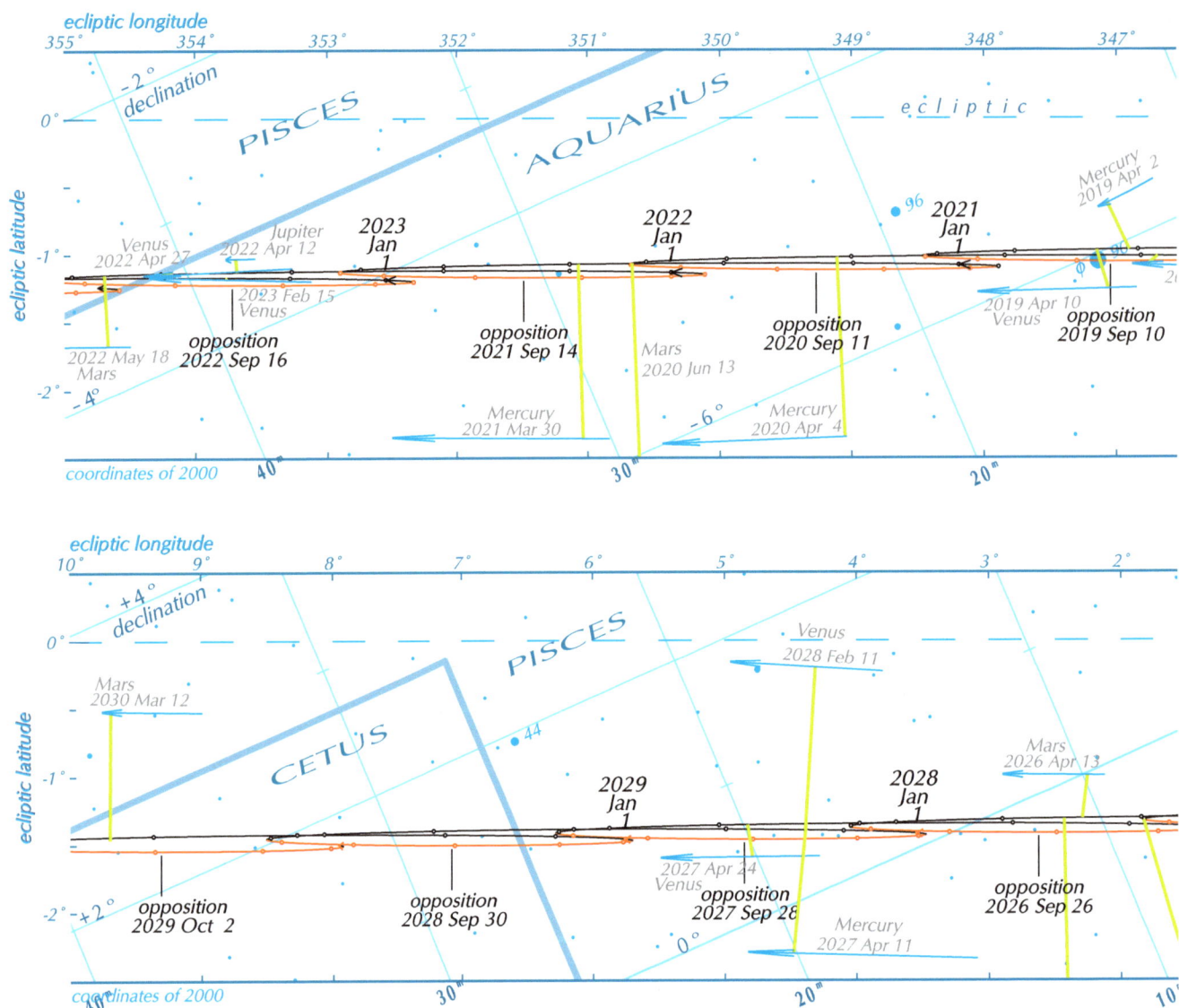

Top chart:

ecliptic longitude

355° 354° 353° 352° 351° 350° 349° 348° 347°

−2° declination

0°

ecliptic

ecliptic latitude

PISCES   AQUARIUS

−1°

Mercury 2019 Apr 2

96 2021 Jan 1

2023 Jan 1   2022 Jan 1

Venus 2022 Apr 27   Jupiter 2022 Apr 12

2023 Feb 15 Venus

opposition 2022 Sep 16

2022 May 18 Mars

opposition 2021 Sep 14

Mars 2020 Jun 13

opposition 2020 Sep 11

2019 Apr 10 Venus

opposition 2019 Sep 10

Mercury 2021 Mar 30   Mercury 2020 Apr 4

−2° −4°

−6°

coordinates of 2000   40ᵐ   30ᵐ   20ᵐ

Bottom chart:

ecliptic longitude

10° 9° 8° 7° 6° 5° 4° 3° 2°

+4° declination

0°

ecliptic latitude

Venus 2028 Feb 11

Mars 2030 Mar 12   PISCES

CETUS

44

Mars 2026 Apr 13

−1°

2029 Jan 1   2028 Jan 1

2027 Apr 24 Venus

opposition 2029 Oct 2   opposition 2028 Sep 30   opposition 2027 Sep 28   opposition 2026 Sep 26

Mercury 2027 Apr 11

0°

−2° +2°

coordinates of 2000   40ᵐ   30ᵐ   20ᵐ   10ᵐ

Neptune has been in the southern celestial hemisphere since 1944. Uranus overtook it in 1993 (an extraordinary triple event as seen from Earth) and is drawing ahead by about 1.77° each year. The path of Uranus in our maps is about 1.85 longer than that of Neptune.

Neptune being in Aquarius, its oppositions happen when the Sun is across the sky in Leo, that is, in September. They move into October in 2029. For observers in north-hemisphere latitudes, Neptune first appears in the pre-dawn sky in March, starts rising before midnight in June, is at its modest highest around opposition, is setting a few hours after sunset at the end of the year.

Neptune is moving slowly inward, from aphelion in 1968 to perihelion in 2042. But since its orbit is not far from circular, the changes are scarcely discernible. The distance from us at opposition decreases by about 0.006 a.u. a year—nearly 900,000 kilometers, which is tiny compared with the enormous total distance of about 4,500,000,000 km. The apparent width of 2.3″, and magnitude of 7.8, scarcely improve.

Neptune is riding south of the ecliptic, and sliding very gradually farther south, because it was at descending node in 2003 and will be at greatest latitude south in 2044.

Since it is at such a distance, and only about a degree south of the ecliptic plane, each apparent retrograde piece of its path is close to—not far south of—the previous forward piece, and when it turns forward again the next piece is between the two previous ones—in short, Neptune's path is difficult to disentangle on the map! The pieces of the path become slightly wider apart by 2030.

The main improvement (for north-hemisphere observers) is the northing: Neptune's declination at opposition climbs from −7°41′ in

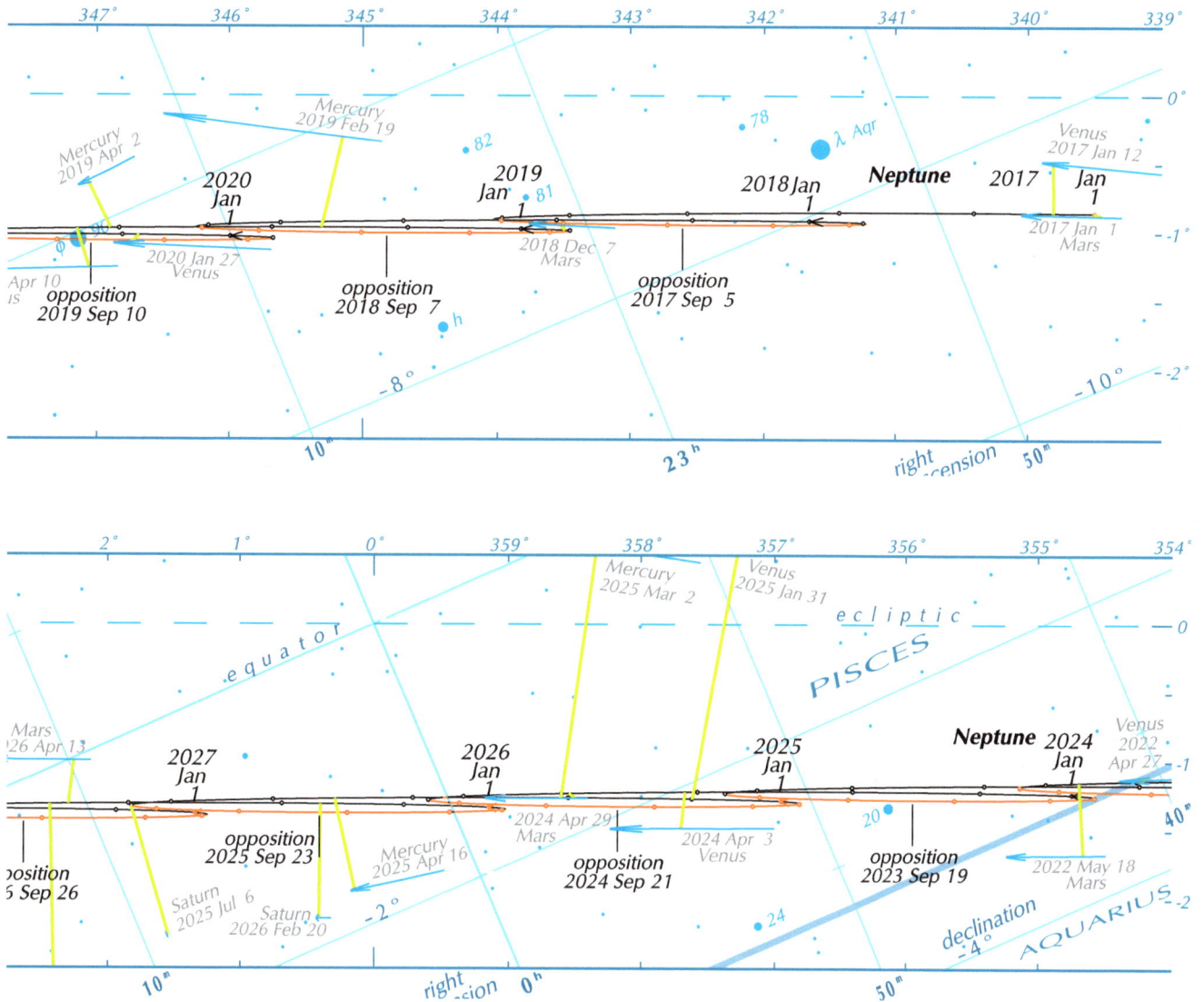

2017, to -0°14´ and +0°35´in 2026 and 2027 (during its back-and-forth crossing of the equator), and +3°56´ in 2030.

The advancing of Neptune from Aquarius into Pisces is, as seen from Earth, a complicated process starting in April 2022, almost at the same moment of a close conjunction with Venus high in the morning sky. Neptune crosses back in August, before its Sep. 16 opposition; starts again forward on Dec. 3; is again closely passed by Venus on 2023 Feb. 15; crosses the border a third time in March; retreats after the opposition of 2023 Sep. 19 and becomes stationary on Dec. 6 almost on the constellation border—a small fraction of a degree short of touching it a fourth time.

And in 2028-2029 comes a crossing into, and in 2029-2030 the crossing back out of, Cetus— the corner of that non-zodiacal constellation that sticks up into Pisces almost as far as the

ecliptic, so that it is often visited by the Moon and planets. The entry into Cetus is a relatively simple three-part event, because the boundary is a north-south one, instead of—like the Aquarius-Pisces border—slanting at a low angle across the planet's route.

For finding dim Neptune, the far more brilliant planets can serve as guides, around times when they are close to it, and are not too near to the Sun. These conjunctions of the planets with Neptune may be helpful:

2017 Jan 1: Mars; elongation 59° from the Sun in the evening sky.
2017 Jan 12: Venus; 47°, evening sky.
2018 Dec 7: Mars; 88°, evening sky.
2022 Apr 27: Venus; 43°, morning sky.
2022 May 18: Mars; 62°, morning sky.
2024 Apr 29: Mars; 40°, morning sky.
2025 Jul 6: Saturn; 102°, morning sky.
2028 Feb 11: Venus; 43°, evening sky.

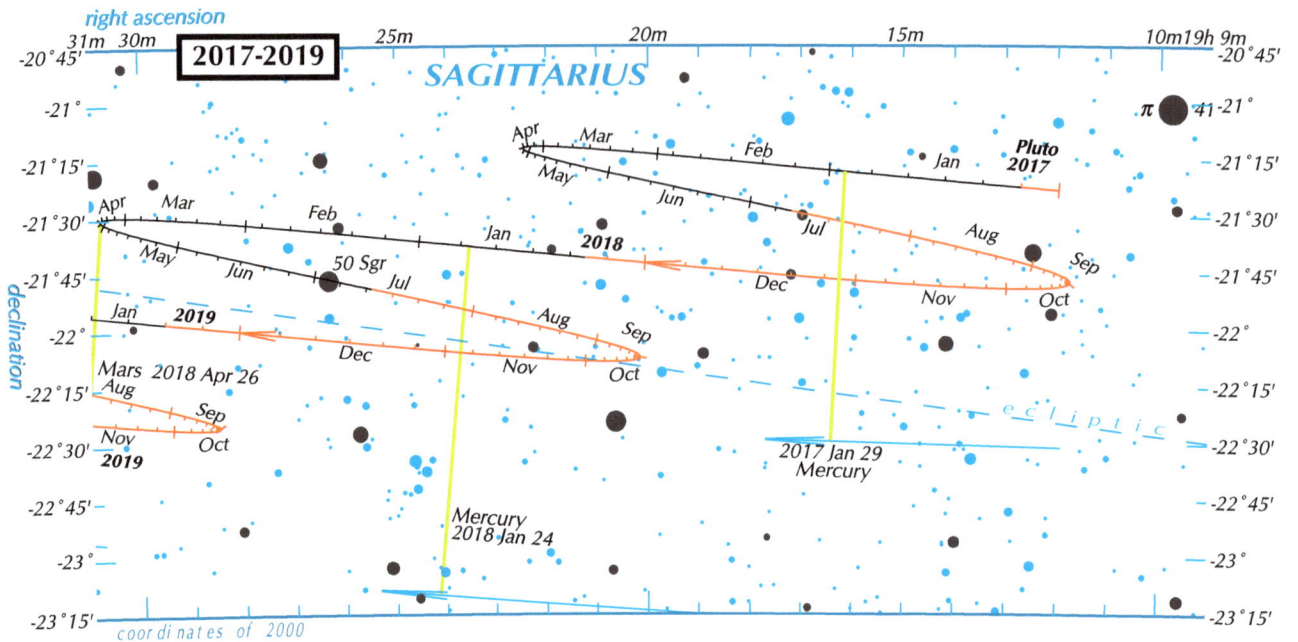

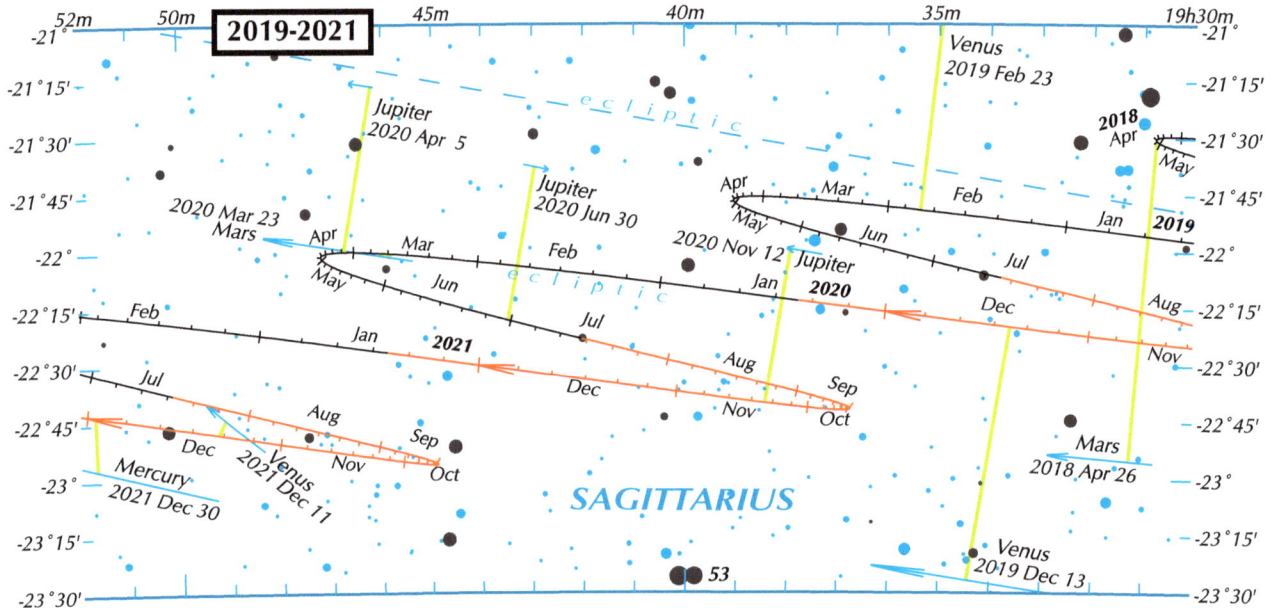

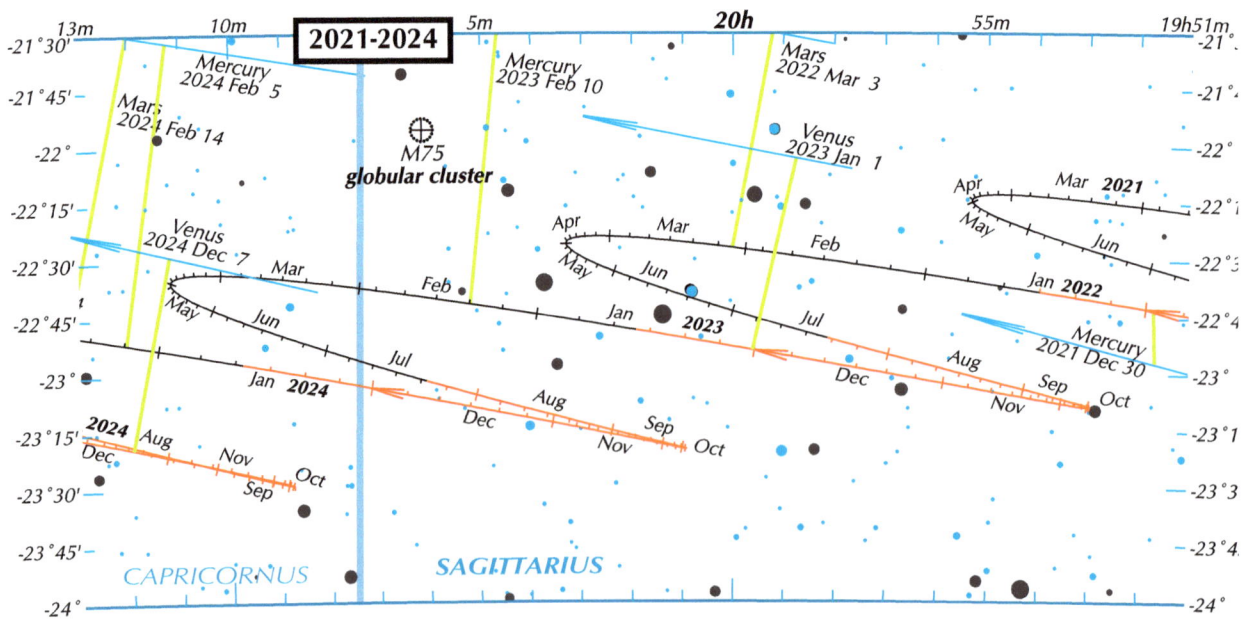

These charts are in the equatorial framework; the ecliptic mode would not save space, and right ascension and declination have to be used in searching for anything as faint as Pluto.

The scale is 3 cm to 1°. The projection is azimuthal equidistant; that is, angular distances are correct from the center of each chart.

Ticks mark Pluto's positions at days 1, 6, 11, 16, 21, and 26 of each month. The track is shown red when it is in the evening sky. Each opposition is in the middle of the retrograde part, at the point where black turns to red (July till 2030, then August). Sun-conjunction is in the middle of the advancing stretch, where red turns to black (in January). These geocentric events falls about 1.8 day later each year.

Appulses with planets are shown if they are closer than 1.5°, and at elongations more than 15° from the Sun.

Plotted in black are stars from the Hipparcos catalogue. Those plotted in blue are from the Tycho catalogue of the Hipparcos mission, which are 9 times more numerous and can be as faint as magnitude 11.5.

Pluto is moving outward, from perihelion in 1989, to 33.4 a.u. from the Sun in 2017, to 36.5 in 2030, toward aphelion in 2114. Its opposition magnitude is 14.2 in 2017, dimming to 14.6 by 2030, so it is dimmer than the faintest stars shown. To find it, use a telescope of aperture 10 inches or more (though it has been seen in 6-inch reflectors); locate the right part of the star field; make a careful sketch of all stars; look on a later night; and find the "star" that has moved.

At opposition on 2018 July 29, Pluto is close to the star 50 Sagittarii, magnitude 5.6.

In 2020, Mars and Jupiter go by in Pluto's foreground, and there is a compound conjunction. Mars passes 0.71° south of Jupiter on March 20, then only 0.01° south of Pluto on March 23; then Jupiter comes on past Pluto, at 0.74° to its north, on April 5. This happens in the morning sky, not long before Pluto is at its stationary point preceding opposition. Because Jupiter is moving back and forth more rapidly, the two have appulses again on June 30 and Nov. 12.

In 2022 and 2023, Pluto comes within a degree of the dense globular cluster M75 (75 in the Messier list of deep-sky objects), 67,500 light-years distant.

Pluto descends through the ecliptic on 2018 Oct. 25. After that, the S-shape of its forth-and-back apparent path grows unsymmetric, the backward extreme thinner, till in 2023 it is virtually a point. Then in 2024 the path crosses over itself, becoming looped; for the first time the August part lies just south of the November part,

Pluto ventures from Sagittarius into Capricornus in the last hours of 2023 Feb. 28. From revolving Earth, it seems to retreat into Sagittarius in July; finally moves into Capricornus on 2024 Jan. 2.

You can see in the last chart that the southward sag of Pluto's trajectory flattens out. On 2030 Oct. 10 (just before an Oct. 23 stationary point) Pluto reaches a southern extreme of declination, 23°54′.

Oligarchs and Plutocrats

News media constantly refer to Russian "oligarchs," such as Anatoly Chubais, Boris Berezovsky, Mikhail Khodorkovsky, Oleg Deripaska, Roman Abramovich. What they mean is "plutocrats." Even journals as literate as the *New Yorker* and *New York Times* seem stuck with this popular misnomer.

Oil was among the assets that, when the Soviet state was privatized, were grabbed by these men, so there may be a notion that "oligarch" has something to do with oil. The Greek roots of these words are *olig-*, "few," *plout-*, "wealth," *krat-*, "power," *arkh-*, "rule."

Oligarchs were members of small groups that ruled some ancient Greek cities; for instance the "Four Hundred" who briefly overthrew Athenian democracy in 411 B.C. A Russian who rakes in a lot of money in connivance with corrupt officials is a plutocrat; he is not an oligarch, especially if he's fallen out with the regime and been exiled or imprisoned. These men have also been called kleptocrats, from another Greek root, *klept-*, "steal."

The Greek historian Plutarch (46-120 A.D.), author of the *Parallel Lives* and of philosophical essays, was not really a "wealth-ruler" despite his name.

2024-2027

CAPRICORNUS

34m · 30m · 25m · 20m · 15m · 20h12m

-21°45' · -22° · -22°15' · -22°30' · -22°45' · -23° · -23°15' · -23°30' · -23°45' · -24° · -24°15'

Mars 2024 Feb 14

Pluto 2024 Feb

Apr · Mar · Feb · Jan · **2025** · Dec · Nov · Oct · Sep · Aug · Jul · Jun · May

Apr · Mar · Feb · Jan · **2026** · Jul · Jun · May

Apr · Mar · Feb · Jan · **2027** · Dec · Nov · Oct · Sep · Jul · Jun · May

2026-2030

CAPRICORNUS

55m · 50m · 45m · 40m · 35m · 20h33m

Pluto 2026

Mar · Feb · Jan · **2030** · Jul · Dec · Nov · Oct · Sep · Aug

Apr · Mar · Feb · May · Jun · **2029** · Jan · Jul · Dec · Nov · Oct · Sep · Aug

Apr · Mar · Feb · May · Jun · **2028** · Jan · Jul · Dec · Aug · **2027** · Apr · May · Feb

southernmost declination

2030 Oct

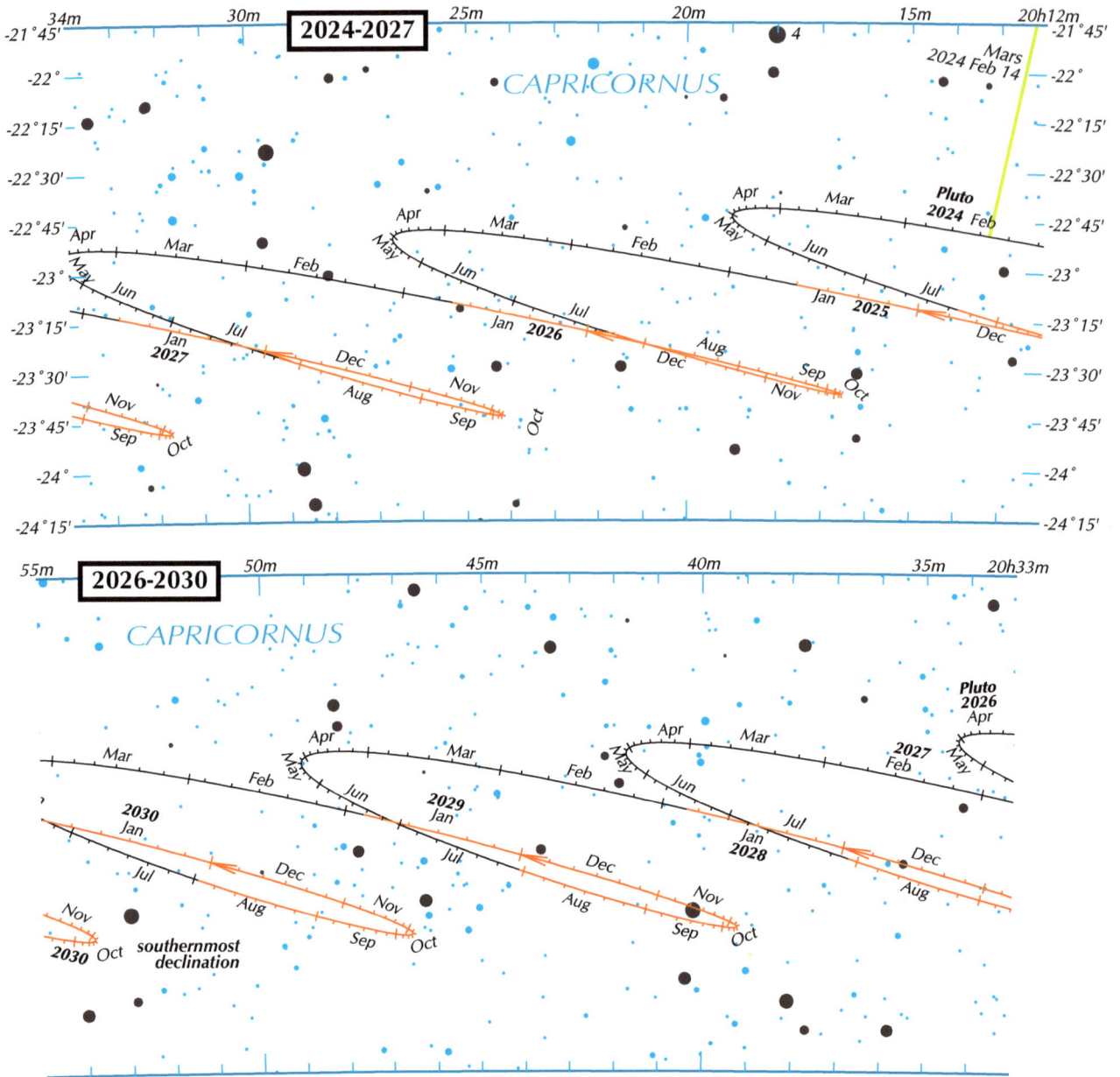

Pluto's future travel through the south and back to where it was discovered.

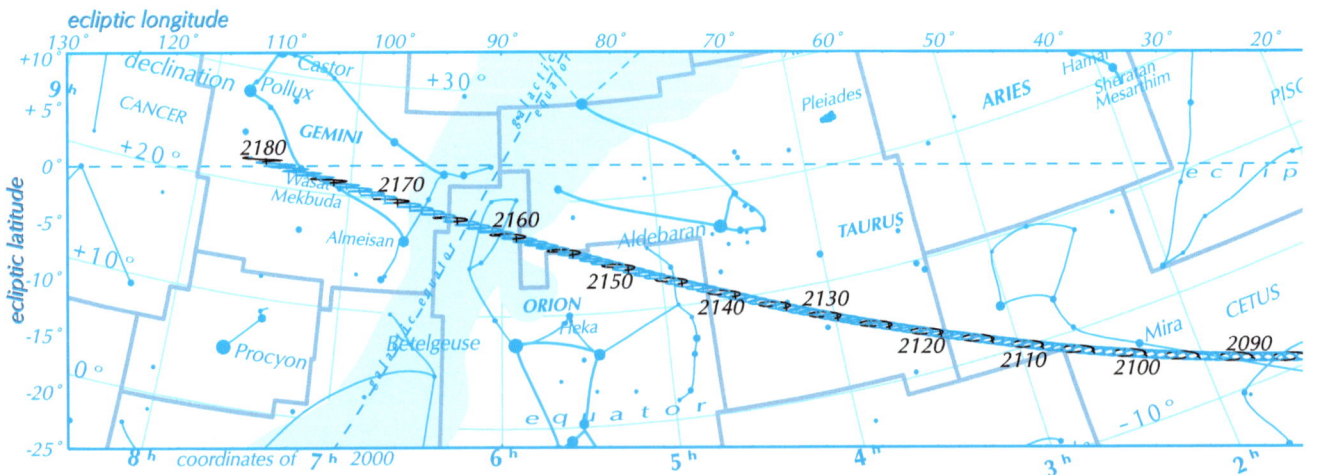

ecliptic longitude

130° · 120° · 110° · 100° · 90° · 80° · 70° · 60° · 50° · 40° · 30° · 20°

declination

+10°

CANCER · Pollux · Castor · GEMINI

+30°

Pleiades · ARIES · Hamal · Sheratan · Mesarthim · PISC

ecliptic latitude

9h · +5° · 0°

+20° · 2180

Wasat · Mekbuda · 2170 · Almeisan · 2160

Aldebaran · TAURUS

+10° · -5°

2150 · 2140 · 2130 · 2120 · 2110 · 2100 · 2090

ORION · Heka · Betelgeuse

0° · -20°

Procyon · Mira · CETUS

equator

-10°

-25° · 8h · coordinates of 7h 2000 · 6h · 5h · 4h · 3h · 2h

Voyages down to Pluto

"Charon's obol" is the coin you put in the mouth of the dead, so that she can pay the ferryman's fare across the River Styx.

Long before New Horizons, there were expeditions by heroes who "harrowed hell"—raided the underworld, through caves or lakes or other of its entrances.

Christ made a *descensus ad inferos* before ascending to heaven.

In Book 11 of the *Odyssey*, Odysseus by sailing as far as the world's western end is able to meet with the spirits of the dead, who give him information useful to the story.

Theseus of Athens and his companion-in-arms Peirithous the Lapith ventured down because Peirithous wanted to steal Persephone from her husband Hades; invited to sit, they found themselves unable to rise. Heracles came down and rescued Theseus by tearing him from the rock—which was why Theseus's Athenian descendants had small buttocks—but the earth shook warningly when Heracles tried to free Peirithous.

When Eurydice died of a snakebite, her husband Orpheus the supreme musician went down and his sweet voice and lyre put Charon and the hell-hound Cerberus to sleep and soft-ened the hearts of Pluto and Persephone and even of the Fates. He was allowed to lead Eurydice away, on condition that he not look back at her till they reached the upper world; he turned when he had done so, but she had not yet reached the portal, and so he lost her.

Admetus was so likeable and hospitable that he had a promise from the gods: he could, when his time came, be spared death if someone would go in his place. His old parents refused, but his loving wife Alcestis volunteered.

katêlthon an, kai m' outh' ho Ploutônos kuôn
outh' houpi kôpê psukhopompos an Kharôn
eskhon, prin es phôs son katastêsai bion.

"I would go down, and neither Pluto's dog
Nor Charon, soul-conductor, with his oar,
Would stop me bringing you to light and life."

—Lines 360-362 of *Alcestis*, staged at Athens in 438 B.C. There is some question whether this early play by Euripides is really a tragedy or is at least partly satirical. Admetus says he would go down to rescue her, if only he had the musical ability of Orpheus He doesn't have to worry: Heracles is again the rescuer, by entering through Alcestis's tomb and wrestling with Thanatos, "Death."

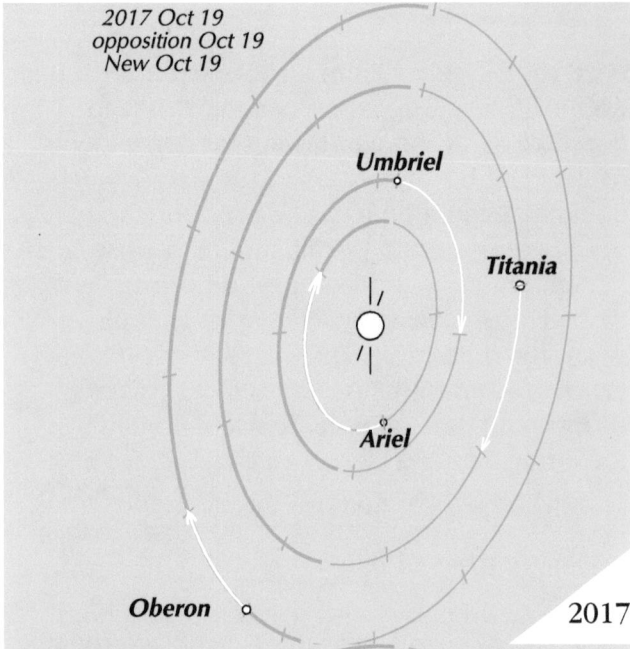

2017 Oct 19
opposition Oct 19
New Oct 19

Umbriel
Titania
Ariel
Oberon

2017

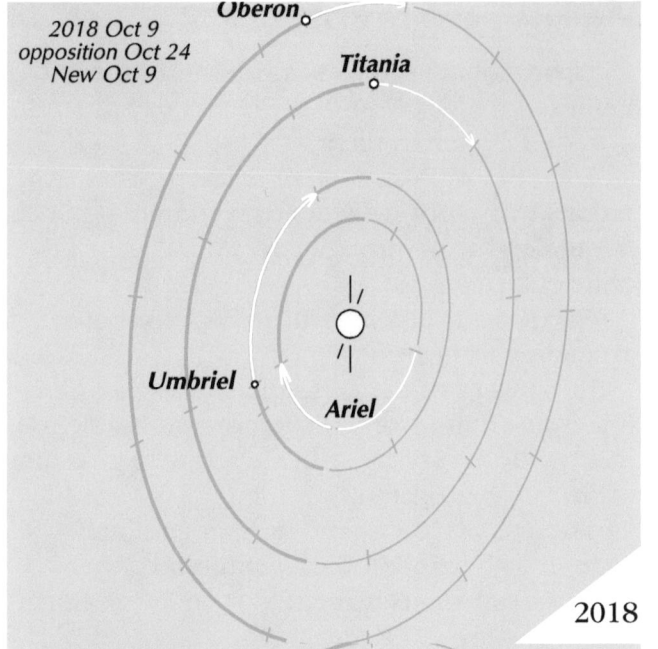

2018 Oct 9
opposition Oct 24
New Oct 9

Oberon
Titania
Umbriel
Ariel

2018

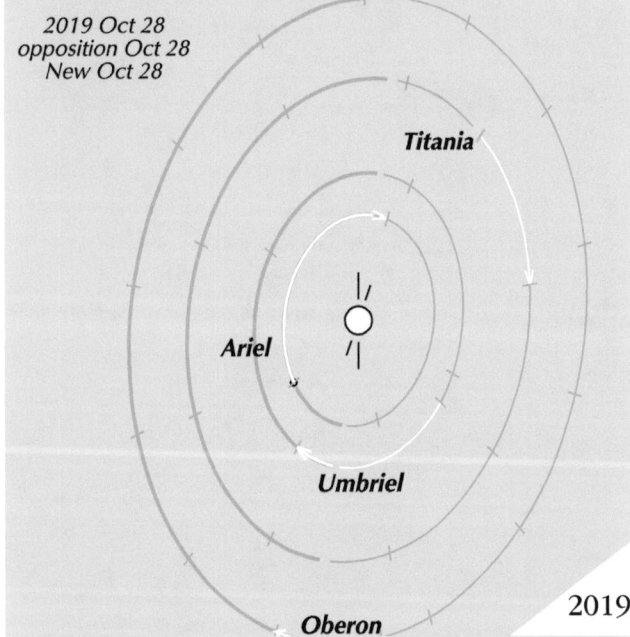

2019 Oct 28
opposition Oct 28
New Oct 28

Titania
Ariel
Umbriel
Oberon

2019

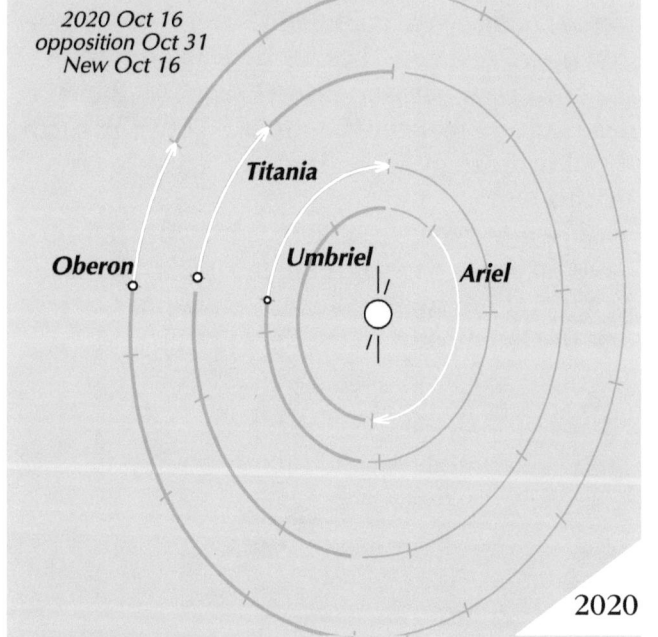

2020 Oct 16
opposition Oct 31
New Oct 16

Titania
Oberon
Umbriel
Ariel

2020

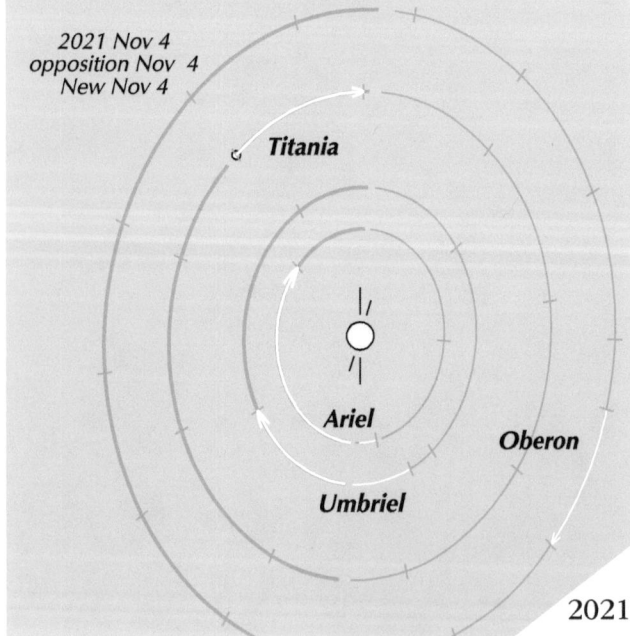

2021 Nov 4
opposition Nov 4
New Nov 4

Titania
Ariel
Oberon
Umbriel

2021

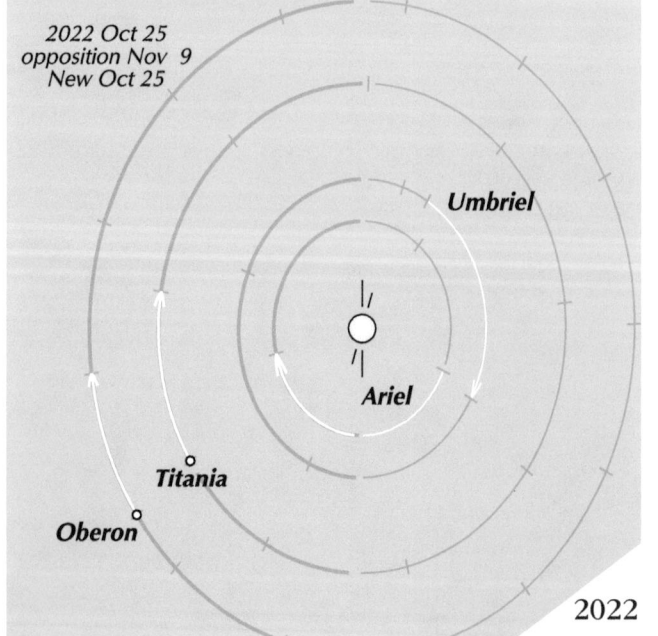

2022 Oct 25
opposition Nov 9
New Oct 25

Umbriel
Ariel
Titania
Oberon

2022

Satellite orbits

←Uranus

Shown here are only the largest and brightest of Uranus's 27 and Neptune's 14 known satellites.

The diagrams illustrate how the satellites move around their planets, but they are only for one selected day in each year. The day chosen is that of the planet's opposition—unless the Moon phase to which this is closest is Full, in which case we choose the day of the preceding New Moon. The label shows the day of the picture, and the day of the nearest Moon phase.

Uranus's oppositions are, in our time, coinciding rather closely with New Moons in even years and Full Moons in odd years. With Neptune the pattern is not so regular: opposition is close to Full Moon in 2017, 2024, 2026, and 2028.

The path of each satellite is shown in white for the day (0h to 24h Universal Time). The rest of the satellite's orbit is in blue, with ticks at the beginning of the following Universal Time days. Thus you can see that, for instance, Ariel's period is about 2.5 days and Oberon's about 13.5.

The scale is 1 millimeter to 1 second of arc. The planets are drawn to scale, but the satellites' sizes are exaggerated by 10.

Celestial north is up. The longer pointers above the planets are to the north celestial pole, the shorter to the north ecliptic pole.

Neptune

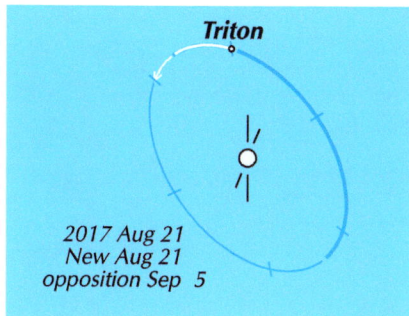

Triton

2017 Aug 21
New Aug 21
opposition Sep 5

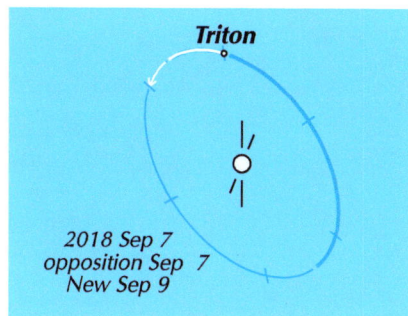

Triton

2018 Sep 7
opposition Sep 7
New Sep 9

2019 Sep 10
opposition Sep 10
First Quarter Sep 6

Triton

Triton

2020 Sep 11
opposition Sep 11
Last Quarter Sep 10

2021 Sep 14
opposition Sep 14
First Quarter Sep 13

Triton

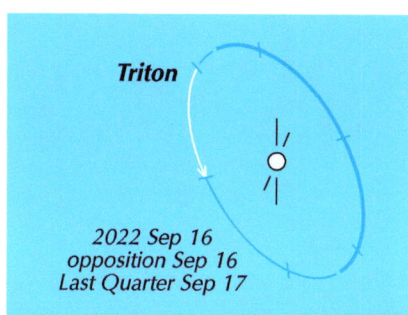

Triton

2022 Sep 16
opposition Sep 16
Last Quarter Sep 17

68

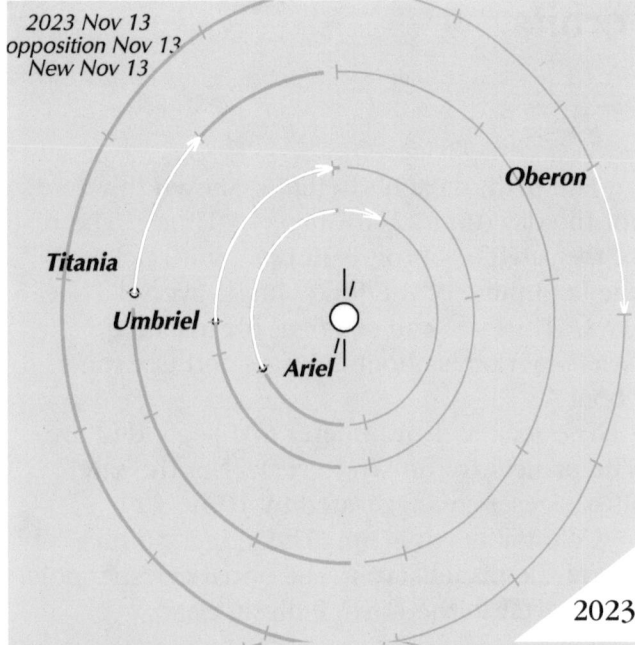

2023 Nov 13
opposition Nov 13
New Nov 13

Oberon

Titania

Umbriel

Ariel

2023

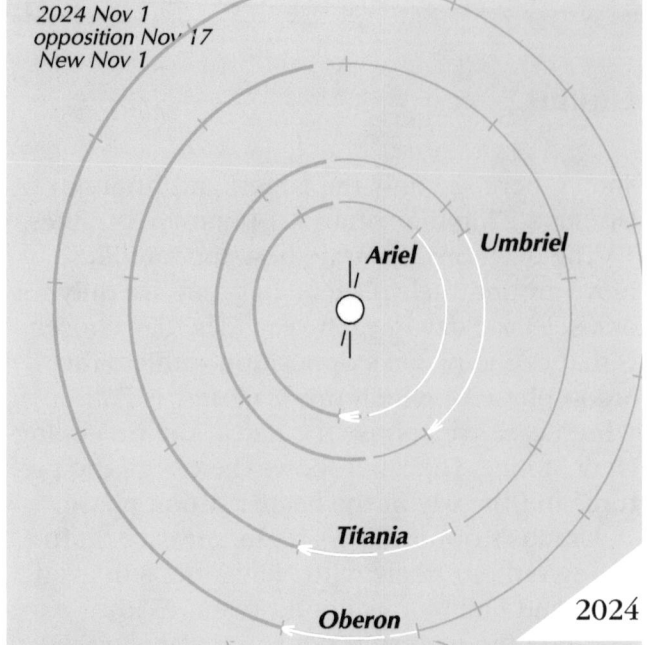

2024 Nov 1
opposition Nov 17
New Nov 1

Ariel

Umbriel

Titania

Oberon

2024

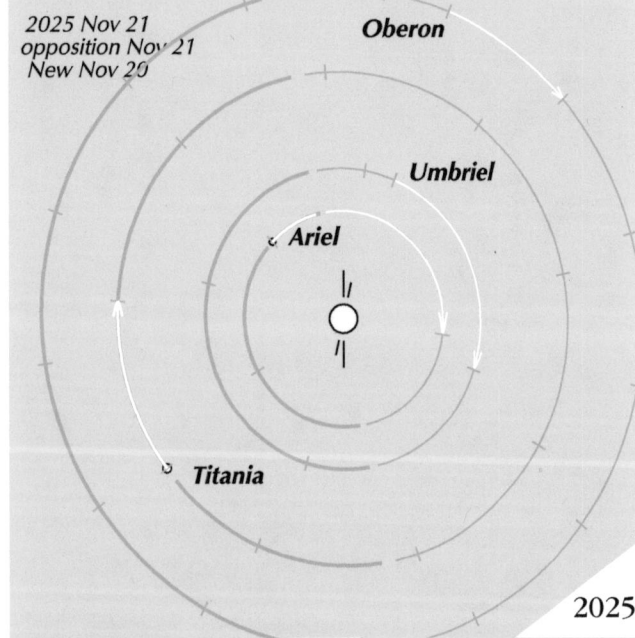

2025 Nov 21
opposition Nov 21
New Nov 20

Oberon

Umbriel

Ariel

Titania

2025

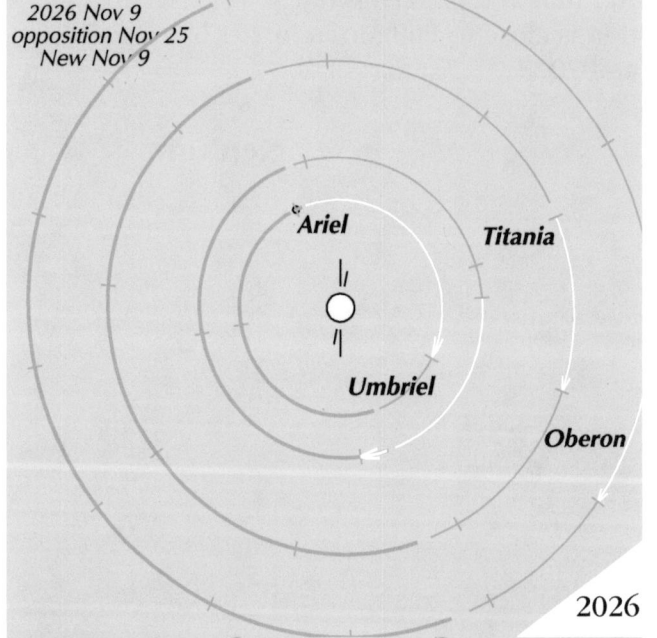

2026 Nov 9
opposition Nov 25
New Nov 9

Ariel

Titania

Umbriel

Oberon

2026

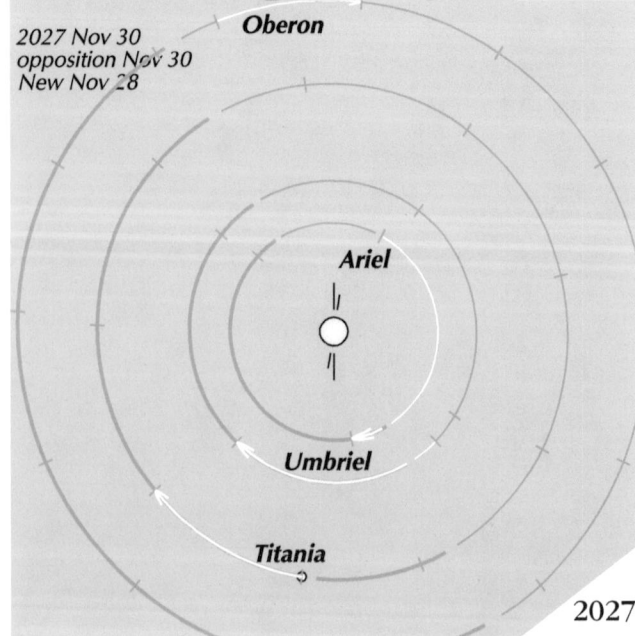

2027 Nov 30
opposition Nov 30
New Nov 28

Oberon

Ariel

Umbriel

Titania

2027

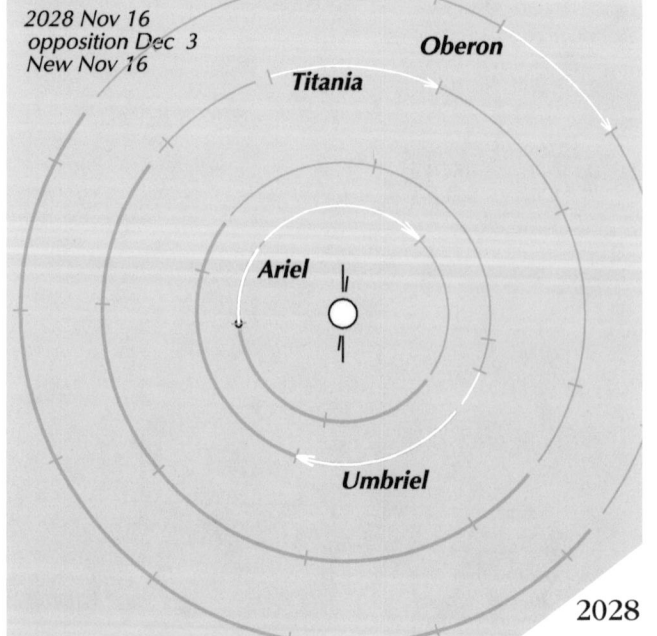

2028 Nov 16
opposition Dec 3
New Nov 16

Titania

Oberon

Ariel

Umbriel

2028

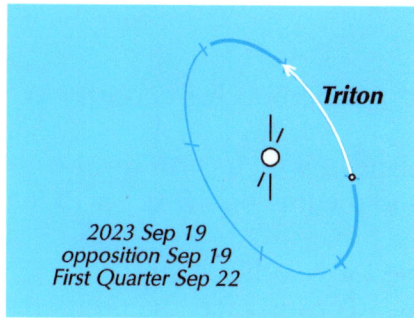

2023 Sep 19
opposition Sep 19
First Quarter Sep 22

2024 Sep 3
opposition Sep 21
New Sep 3

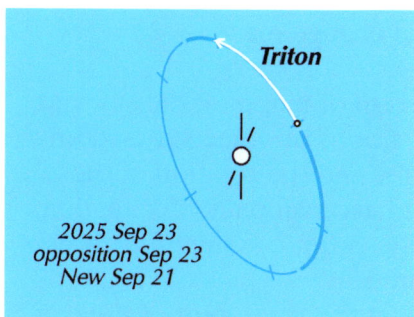

2025 Sep 23
opposition Sep 23
New Sep 21

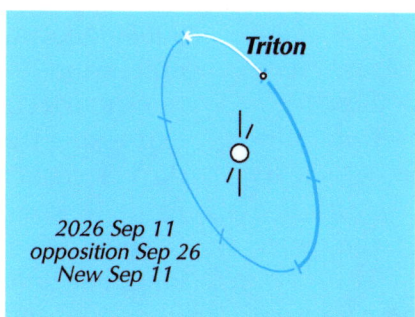

2026 Sep 11
opposition Sep 26
New Sep 11

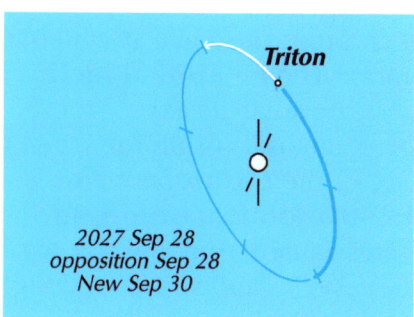

2027 Sep 28
opposition Sep 28
New Sep 30

2028 Sep 18
opposition Sep 30
New Sep 18

Glossary

Here are minimal explanations of terms we have had to use. A much fuller astronomical glossary is our book *Albedo to Zodiac*.

aphelion: the outermost point in an orbit around the Sun.

apparent: the apparent diameter of a planet is the angle it subtends as seen from Earth, as distinct from its physical width in kilometers or other linear measure.

arc: the phrases "minutes of arc" and "seconds of arc" are used to distinguish these angles—parts of the circle of 360°—from minutes and seconds of right ascension, which are fractions of an hour.

astronomical unit (a.u. or AU): the mean distance between Sun and Earth.

ascending node: the point where a planet's orbital plane crosses northward through the ecliptic.

barycenter: the center of mass of two bodies, such as a planet and its satellite; both revolve around it.

celestial equator: the rotational plane of the Earth; the series of points in the sky above Earth's equator.

celestial sphere: an imaginary sphere centered Earth; stars and constellations can be mapped onto it.

conjunction with the Sun: instant when a planet is in the same direction as the Sun, that is, has the same longitude, though it is usually north or south of the Sun. For Mercury and Venus, **inferior conjunction** is when they pass on the Sun's nearer side, and **superior conjunction** is on the farther side.

declination: angular distance north or south of the celestial equator. It corresponds to latitude on the Earth.

descending node: the point where a planet's orbital plane crosses southward through the ecliptic.

direct motion: motion eastward against the starry background, the normal motion of planets in the solar system. Similarly, **direct rotation** is rotation such that the sky appears to move westward. Another word for the same thing is **prograde**; and the opposite kind of motion or rotation is **retrograde**.

eccentricity of an orbit: its departure from circularity. A circle has eccentricity 0; ellipses have eccentricity between 0 and 1; a parabola

has eccentricity 1; a hyperbola has eccentricity more than 1.

ecliptic: the plane in which Earth revolves around the Sun; and the line on the sky marking this plane.

epoch: the time for which something is intended to be true. For instance, the position (right ascension and declination) of a planet in 2020 may be given for the standard epoch of 2000, ignoring the change made by 20 years of precession.

equatorial system of mapping the sky and expressing the positions of objects: based on the celestial equator and the rotation of the Earth. It uses right ascension and declination, in contrast with the ecliptic system which uses longitude and latitude.

geocentric: from the viewpoint of the center of the Earth.

heliocentric: centered on the Sun.

inclination of an orbit: the angle at which it intersects the ecliptic. The **rotational inclination** of a planet is the angle its axis of rotation makes withthe plane of its orbit; for instance, in the case of the Earth, about 23°.

International Astronomical Union: an official body of professional astronomers; it for instance records discoveries and approves designations.

invariant plane of the solar system: the general plane of its revolving matter, determined mainly by Jupiter. It differs by 2.65° from the ecliptic plane.

Jovian: of Jupiter (an alternative form of whose name is *Jove*).

latitude: angular distance north and south of a plane; usually, in astronomy, the plane of the ecliptic.

Kepler's laws: (1) A planet's orbit is an ellipse, with the Sun at one of the two foci. (2) A line from the planet to the Sun sweeps out an equal area in an equal time. (3) The orbital period (in years) squared equals the semi-major axis (average distance, in astronomical units) cubed.

limb: the visible edge of a spherical body, such as the Sun or Moon.

longitude:' angular distance around, parallel to the ecliptic.

magnitude: the astronomical system for expressing brightness. Very bright stars have magnitude 1 (a few, 0 or negative); stars just visible to the naked eye have magnitude around 6; Pluto is as dim as 14. For historical reasons, the scale is logarithmic and each step of 1 represent a difference of about 2.5 in amount of light—to be precise, the factor is the 5th root of 100, or 2.5119 . . .

Messier objects (M1, M2, etc.): in Charles Messier's catalogue (1771) of deep-sky objects (clusters, nebulae, galaxies) that could be mistaken for comets.

minute or angle, or "of arc," is 1/60 of a degree; but a minute of right ascension is 1/60 of an hour, therefore 15/60 along the equator but smaller if nearer to the poles.

node: see **ascending** and **descending node**.

occultation: an occasion when a body hides another; for instance the Moon occults planets or stars, and a planet may occult its satellites.

opposition: the instant when a planet or other moving body is in the opposite direction from the Sun, that is, is at a longitude 180° from the Sun's. It is the center of the best time for observing the planet, which is highest at midnight.

perihelion: the innermost point in an orbit around the Sun.

planetesimal: a presumed clump of matter which, by joining with others, went toward the forming of a planet.

precession: gradual change in the direction of a body's axis of rotation. For instance, Earth's north pole sweeps a large circle in the sky, over a period of 25,800 years. Therefore the celestial equator has shifted. Since our sky maps are based on the vernal equinox point, where the equator crosses the ecliptic, all stars gradually change their map positions. Maps are now based on the standard epoch of 2000.

prograde: see **direct**.

primary: the large body around which a satellite revolves.

retrograde: the opposite of **direct** motion or rotation.

right ascension: the astronomical way of measuring angular distance around the celestial sphere. It starts from the **vernal equinox point** and is measured eastward, either in hours (24 of them around the sky) or in degrees (360 of them). Thus the star Betelgeuse is, roughly, at R.A. 6 hours, or 90°.

second or angle, or "of arc," is 1/3600 of a degree; but a second of right ascension is 1/3600 of an hour, therefore 15/3600 along the equator but smaller if nearer to the poles.

semi-major axis of an ellipse: half of its axis, that is, of its longer dimension. In a planet's elliptical orbit, it is equal to the average distance from the Sun (at one focus of the orbit) to the planet

sidereal: of the stars (Latin *sidus*, "star," plural *sidera*).

sidereal period: the time taken by a planet to go once around the sky, as seen from the Sun, against the starry background.

synodic period: the time from one event of a planet as seen from Earth, such as its opposition, to the next such event. For instance, Uranus's synodic period of 370 days means that Earth, after passing Uranus at opposition, takes a year and then roughly 5 more days to catch up with Uranus again.

syzygy: an event in which Earth, Moon, and Sun are in a line; i.e. a New or Full Moon.

terrestrial: of the Earth (Latin *terra*; but the four inner planets— Mercury, Venus, Earth, Mars—are classed as the "terrestrial" or Earth-like ones in contrast with the Jovian giants.

transit: passage of a small body in front of a larger one; for instance Venus across the face of the Sun, or Jupiter's satellites across it.

vernal equinox direction: the direction toward the point where the ecliptic cuts northward through the celestial equator. The Sun is at this point at the March equinox, which marks spring (Latin *ver*) for Earth's northern hemisphere. This point is used as the origin, or zero point, for mapping the sky. Over the centuries it shifts westward because of precession.

zodiac: the band of constellations around the sky through which the Sun, Moon, and major planets pass. There are traditionally 12 zodiacal constellations (Pisces, Aries, Taurus, Gemini, Cancer, Leo, Virgo, Libra, Scorpius, Sagittarius, Capricornus, Aquarius), but the ecliptic also passes through Ophiuchus, and the parts of some other constellations come near enough that the Moon and planets sometimes are in them.

Tables

Some sizes and distances

In the first column are minor planet numbers, or, for satellites, the primary and the number in its satellite system. The third column gives the year of discovery. "Distance" means average distance from the Sun, except for satellites, for which it is from the center of the primary. "Diameter" means generally the equatorial diameter. Figures are given to low precision. Some may be changed by later studies. "DP" means "dwarf planet.

			diameter km	distance km	AU	
Sun			1,392,000	0	0	
Mercury			4,880	57,900,000	0.387	
Venus			12,100	108,208,000	0.723	
Earth			12,756	149,698,000	1	
Mars			6,790	227,900,000	1.524	
Jupiter			143,000	778,600,000	5.204	
Saturn			120,500	1,433,500,000	9.683	
Uranus		1781	51,100	2,875,000,000	19.218	
Neptune		1846	49,500	4,504,400,000	30.110	

transnepunians

134340	Pluto	1930	2,380	5,906,380,000	39.48	DP
20000	Varuna	2000	670	6,425,000,000	42.950	DP?
28978	Ixion	2001	800	5,933,000,000	39.664	DP?
50000	Quaoar	2002	1,100	6,525,000,000	43.616	DP?
90377	Sedna	2003	1,000	75,800,000,000	506.8	DP?
90482	Orcus	2004	917	5,894,000,000	39.398	DP?
136108	Haumea	2004	2,000	6,465,000,000	43.218	DP
136199	Eris	2005	2,326	10,166,000,000	67.781	DP
136472	Makemake	2005	1,478	6,839,000,000	45.715	DP

satellites

Earth	Moon		3,476	384,400	
Jupiter 1	Io	1610	3,644	421,700	
2	Europa	1610	3,122	670,900	
3	Ganymede	1610	5,268	1,070,400	
4	Callisto	1610	4,820	1,883,000	
Saturn 6	Titan	1655	5,151	1,221,870	
Neptune 1	Triton	1846	2,706	354,800	
Pluto 1	Charon	1978	1,212	19,590	

main-belt asteroids

1	Ceres	1801	962	414,010,000	2.767	DP
2	Pallas	1802	524	414,700,000	2.772	
3	Juno	1804	274	399,400,000	2.67	
4	Vesta	1807	512	353,300,000	2.362	

Orbital characteristics

q: perihelion distance, in astronomical units.
a: mean distance ("semimajor axis").
Q: aphelion distance.
P: sidereal period, in years.
p: synodic period, in days.
e: eccentricity.
i: inclination to the ecliptic, in degrees.

name	q AU	a AU	Q AU	P years	p days	e	i °
Uranus	18.33	19.22	20.11	84.02	369.66	0.05	0.77
Neptune	29.81	30.11	30.33	164.8	367.49	0.009	1.77
Pluto	29.66	39.48	49.31	248	366.73	0.249	17.16

Oppositions

Column 1: Julian Date. Other columns are: right ascension, declination (both for epoch 2000); heliocentric distance, longitude, and latitude; geocentric distance; apparent magnitude; apparent diameter in seconds.

	Uranus	RA (2000)	decl	hedis	helon	helat	gedis	mag	dia"
2458046.223	2017 Oct 19 17	1 38 24	9 35	19.911	26.52	-0.57	18.915	5.7	3.7
2458415.523	2018 Oct 24 1	1 53 37	11 3	19.870	30.56	-0.53	18.875	5.7	3.7
2458784.835	2019 Oct 28 8	2 9 3	12 29	19.827	34.61	-0.49	18.833	5.7	3.7
2459154.152	2020 Oct 31 16	2 24 45	13 51	19.780	38.68	-0.45	18.788	5.7	3.7
2459523.489	2021 Nov 4 24	2 40 43	15 11	19.731	42.78	-0.40	18.739	5.6	3.7
2459892.844	2022 Nov 9 8	2 56 59	16 26	19.678	46.91	-0.35	18.687	5.6	3.7
2460262.214	2023 Nov 13 17	3 13 34	17 38	19.621	51.06	-0.30	18.632	5.6	3.8
2460631.603	2024 Nov 17 2	3 30 28	18 44	19.561	55.24	-0.25	18.572	5.6	3.8
2461001.007	2025 Nov 21 12	3 47 41	19 45	19.497	59.44	-0.20	18.509	5.6	3.8
2461370.436	2026 Nov 25 22	4 5 13	20 40	19.431	63.67	-0.14	18.444	5.6	3.8
2461739.881	2027 Nov 30 9	4 23 4	21 29	19.363	67.92	-0.08	18.377	5.6	3.8
2462109.344	2028 Dec 3 20	4 41 13	22 10	19.294	72.20	-0.03	18.308	5.6	3.8
2462478.833	2029 Dec 8 8	4 59 38	22 45	19.224	76.51	0.03	18.239	5.5	3.8
2462848.350	2030 Dec 12 20	5 18 20	23 11	19.155	80.85	0.09	18.171	5.5	3.9
2463217.884	2031 Dec 17 9	5 37 16	23 29	19.087	85.22	0.15	18.102	5.5	3.9

	Neptune	RA (2000)	decl	hedis	helon	helat	gedis	mag	dia"
2458001.717	2017 Sep 5 5	22 57 20	-7 41	29.947	342.85	-0.91	28.939	7.8	2.3
2458369.260	2018 Sep 7 18	23 5 40	-6 53	29.940	345.08	-0.97	28.933	7.8	2.3
2458736.800	2019 Sep 10 7	23 13 59	-6 4	29.935	347.31	-1.02	28.928	7.8	2.3
2459104.343	2020 Sep 11 20	23 22 18	-5 15	29.929	349.55	-1.08	28.922	7.8	2.3
2459471.881	2021 Sep 14 9	23 30 36	-4 26	29.922	351.79	-1.13	28.917	7.8	2.3
2459839.423	2022 Sep 16 22	23 38 53	-3 36	29.915	354.03	-1.18	28.910	7.8	2.3
2460206.962	2023 Sep 19 11	23 47 9	-2 46	29.906	356.27	-1.23	28.902	7.8	2.3
2460574.502	2024 Sep 21 0	23 55 25	-1 56	29.897	358.51	-1.28	28.893	7.8	2.3
2460942.028	2025 Sep 23 13	0 3 40	-1 5	29.887	0.75	-1.33	28.884	7.8	2.3
2461309.559	2026 Sep 26 1	0 11 54	0-14	29.878	2.99	-1.37	28.876	7.8	2.3
2461677.089	2027 Sep 28 14	0 20 9	0 35	29.870	5.23	-1.41	28.868	7.8	2.3
2462044.607	2028 Sep 30 3	0 28 23	1 26	29.863	7.47	-1.45	28.862	7.8	2.3
2462412.132	2029 Oct 2 15	0 36 38	2 16	29.857	9.71	-1.49	28.856	7.8	2.3
2462779.647	2030 Oct 5 4	0 44 54	3 6	29.852	11.95	-1.53	28.853	7.8	2.3
2463147.172	2031 Oct 7 16	0 53 11	3 56	29.849	14.20	-1.56	28.850	7.8	2.3

	Pluto	RA (2000)	decl	hedis	helon	helat	gedis	g mag	dia"
2457944.421	2017 Jul 9 22	19 17 5	-21 28	33.364	287.88	0.74	32.347	14.2	0.1
2458311.636	2018 Jul 12 3	19 25 27	-21 48	33.600	289.76	0.17	32.583	14.2	0.1
2458678.821	2019 Jul 14 8	19 33 45	-22 6	33.839	291.62	-0.41	32.823	14.2	0.1
2459045.988	2020 Jul 15 12	19 41 59	-22 22	34.080	293.45	-0.97	33.064	14.3	0.1
2459413.120	2021 Jul 17 15	19 50 7	-22 37	34.323	295.26	-1.52	33.307	14.3	0.1
2459780.226	2022 Jul 19 17	19 58 9	-22 49	34.567	297.04	-2.07	33.552	14.3	0.1
2460147.305	2023 Jul 21 19	20 6 6	-23 0	34.813	298.79	-2.60	33.798	14.4	0.1
2460514.362	2024 Jul 22 21	20 13 56	-23 10	35.061	300.52	-3.13	34.047	14.4	0.1
2460881.385	2025 Jul 24 21	20 21 40	-23 18	35.312	302.22	-3.64	34.298	14.4	0.1
2461248.386	2026 Jul 26 21	20 29 18	-23 24	35.565	303.90	-4.15	34.552	14.4	0.1
2461615.366	2027 Jul 28 21	20 36 50	-23 29	35.820	305.56	-4.64	34.808	14.5	0.1
2461982.319	2028 Jul 29 20	20 44 16	-23 33	36.078	307.20	-5.12	35.067	14.5	0.1
2462349.258	2029 Jul 31 18	20 51 36	-23 35	36.337	308.82	-5.59	35.327	14.5	0.1
2462716.163	2030 Aug 2 16	20 58 51	-23 36	36.596	310.42	-6.05	35.587	14.6	0.1
2463083.061	2031 Aug 4 13	21 5 59	-23 35	36.856	312.00	-6.50	35.848	14.6	0.1

www.ingramcontent.com/pod-product-compliance
Lightning Source LLC
Chambersburg PA
CBHW052053190326
41519CB00002BA/207